Twins and Recursion in Digital, Literary and Visual Cultures

Explorations in Science and Literature

Series Editors:

John Holmes, Anton Kirchhofer and Janine Rogers

Explorations in Science and Literature considers the significance of literature from within a scientific worldview and brings the insights of literary study to bear on current science. Ranging across scientific disciplines, literary concepts, and different times and cultures, volumes in this series will show how literature and science, including medicine and technology, are intricately connected, and how they are indispensable to one another in building up our understanding of ourselves and of the world around us.

Published titles

Biofictions, Josie Gill
Imagining Solar Energy, Gregory Lynall
The Diseased Brain and the Failing Mind, Martina Zimmermann
Narrative in the Age of the Genome, Lara Choksey
Writing Remains, Edited by Josie Gill, Catriona McKenzie, Emma Lightfoot
Rereading Darwin's Origin of Species, Richard G. Delisle and James Tierney

Forthcoming titles

The Social Dinosaur, Will Tattersdill
Physics and the Modernist Avant-Garde, Rachel Fountain Eames

Twins and Recursion in Digital, Literary and Visual Cultures

Edward King

BLOOMSBURY ACADEMIC
LONDON • NEW YORK • OXFORD • NEW DELHI • SYDNEY

BLOOMSBURY ACADEMIC
Bloomsbury Publishing Plc
50 Bedford Square, London, WC1B 3DP, UK
1385 Broadway, New York, NY10018, USA
29 Earlsfort Terrace, Dublin 2, Ireland

BLOOMSBURY, BLOOMSBURY ACADEMIC and the Diana logo
are trademarks of Bloomsbury Publishing Plc

First published in Great Britain 2022
This paperback edition published 2023

Copyright © Edward King, 2022

Edward King has asserted his right under the Copyright, Designs
and Patents Act, 1988, to be identified as Author of this work.

For legal purposes the Acknowledgements on p. ix constitute
an extension of this copyright page.

Cover design: Rebecca Heselton
Cover image: Magnia/shutterstock

All rights reserved. No part of this publication may be reproduced or transmitted
in any form or by any means, electronic or mechanical, including photocopying,
recording, or any information storage or retrieval system, without prior
permission in writing from the publishers.

Bloomsbury Publishing Plc does not have any control over, or responsibility for, any
third-party websites referred to or in this book. All internet addresses given in this
book were correct at the time of going to press. The author and publisher regret
any inconvenience caused if addresses have changed or sites have ceased
to exist, but can accept no responsibility for any such changes.

A catalogue record for this book is available from the British Library.

A catalog record for this book is available from the Library of Congress.

ISBN: HB: 978-1-3501-6915-9
PB: 978-1-3503-2307-0
ePDF: 978-1-3501-6916-6
eBook: 978-1-3501-6917-3

Series: Explorations in Science and Literature

Typeset by Integra Software Services Pvt. Ltd.

To find out more about our authors and books visit www.bloomsbury.com
and sign up for our newsletters.

*For Tamara,
Lila, and Margot*

Contents

List of Figures	viii
Acknowledgements	ix
Introduction: Entwined being	1
1 Telepathic twins and the mythotechnesis of cybernetics	31
2 Twins as weird media	73
3 Twins in the Anthropocene	111
4 Twinning in black futurism	139
5 Twin faces as glitches in algorithmic image cultures	169
Conclusion: From digital twins to glitch twins	197
Works Cited	202
Index	215

Figures

2.1 Oswald and Oliver Deuce make themselves the objects of their own experiment in *A Zed & Two Noughts* directed by Peter Greenaway, BFI 1985. All rights reserved 101
4.1 'Aljana Moons III' by Alexis Peskine, 2015. Reproduced with the permission of October Gallery 140
5.1 Grayson and Ethan Dolan trick Apple's Face ID security system in 'Twins Vs. iPhone X Face ID' by Dolan Twins, YouTube, 4 November 2017. Available online: https://www.youtube.com/watch?v=GFtOaupYxq4 (accessed 7 July 2021) 172

Acknowledgements

I can trace the origins of this book back to one morning in March 2015 when there was not one but two little heartbeats on the ultrasound scan. From that moment on, I saw twins everywhere. I saw them in pushchairs or jogging synchronously down the road. They were in every book I read and every film I watched. How could I have not noticed them before? And so began my research on twins, the stories we tell about them and what these stories say about how we connect with one another in an increasingly technological world. My interest eventually led to this contribution to a micro-genre of books on twins, the authors of which are parents of twins, such as Elizabeth A. Stewart and Juliana de Nooy, whose works, like mine, hold in tension the intimacies of parental love with the critical distance of scholarly analysis.

I am enormously grateful for the support I have received from colleagues at the School of Modern Languages at the University of Bristol and the many friends and scholars who have shaped this work through their comments and feedback. I am especially thankful to Joanna Page, Matthew Brown, Jo Crow, Rebecca Kosick, Paul Merchant, Rachel Randall, Naomi Millner, Lucy Bollington, Annie Ring, Zara Dinnen, Will Viney, Tim Martin, Marcelo Buzato and his students at the Universidade Estadual de Campinas. I am thankful to the organizers and attendees of a number of conferences that helped to shape my ideas along the way, including 'Glitches & Ghosts' at Lancaster University, 'Digital Resistance' at the Centre for Digital Culture, King's College London, 'Art in the Anthropocene' at Trinity Dublin and 'Imagin(in)ing the Anthropocene' at the University of the West of England. A special mention goes to all those people who approached me after conference papers and told me their stories about twins. I would also like to thank my anonymous reviewers for their invaluable feedback, as well as Lucy Brown, Lucy Cope and Ben Doyle at Bloomsbury for their support in seeing this book to print.

My biggest debt of gratitude goes to my family: my daughters for being the direct inspiration for the book and a constant source of joy; my wife for her insights on reading draft after draft; and, finally, my parents John and Mari-Carmen for making it all possible.

Introduction: Entwined being

The 2015 documentary *Twinsters* tells the story of identical twin sisters separated at birth who reconnect after finding each other on the internet. The film starts with one of the siblings, LA-based actress Samantha Futterman (co-director of the film with Ryan Miyamoto), telling the story of how the reconnection came about. Samantha received a message on Twitter telling her to check her Facebook account for a message. When she checked, she found a friend request from a French girl studying fashion in London called Anaïs Bordier whose profile picture looked uncannily like her own. 'Oh, that's ... wait, that's a picture of me.' Samantha accepted Anaïs's friend request and they began a dialogue, establishing that they were both born on the same day in South Korea and were subsequently adopted. And so began their journey of mutual discovery. In many ways, *Twinsters* is a twist on the well-worn reunited twins narrative that precedes the digital age by some time. The film acknowledges this at an early stage in an intertextual manner when Anaïs, in her first message to Samantha, expresses her hesitation by saying that she 'doesn't want to be too Lindsay Lohan', a reference to the 1998 remake of the 1961 Hollywood classic *Parent Trap*, itself just a recent iteration of narrative traditions about separated and reuniting twins that can be traced back to ancient myth.

Stories of separated twins being reunited hold a broad cross-cultural and trans-historical appeal. However, each cultural moment adapts these narrative templates about twins to meet its own needs. In the case of *Twinsters*, the reunited twins story is mobilized as a fantasy of social media connectivity. The film insists on the decisive role played by social media in the way the events unfolded. Not only does Anaïs first learn about Samantha's existence after watching a video of her on YouTube but all the initial stages of the dialogue are also mediated by the dominant interfaces of social media. As we listen to Samantha's narrative of their encounter we are shown close-ups of her Facebook page as she accepts her sister's request and reads her first message. In *Twinsters*, the context of social media is not the incidental backdrop to a story about twins discovering ea other after a childhood apart. Rather, the reuniting twins narrative is employed to explore the uncanny nature of constructing identities online. Finding your genetic double on the internet literalizes the feeling that your online identity is both intimately familiar yet somehow other and alien. It also captures the uncanniness induced by the increasing difficulty of separating out what is you and what is not you – what is internal and external to the self – in a world in which experiences are so often circulated through public networks of display and informed by the narrative models provided by mass media.

In an article published in *The Atlantic* in 2014, Alissa Wilkinson drew attention to the sudden flood of films about *doppelgängers*. The year 2013 saw the release of a startling number of films about doubles, including *Enemy* (directed by Denis Villeneuve) and *The Double* (Richard Ayoade). Wilkinson concludes that this surge in interest in doubles is a 'side effect of the digital age'.[1] On the one hand, 'our new interconnectedness' means that finding somebody out there who looks just like you has become extremely easy. Easily accessible software such as that of the website ilooklikeyou.com promises to 'unite the world one face at a time'.[2] On the other hand, the creation of our own doppelgängers has become normalized and part of everyday practice. People routinely construct 'cooler versions' of themselves through their Instagram feeds.[3] The function of twins in narratives of cultures of digital connectivity is often very similar. Like doppelgängers, twins are used to explore the current mutations in our experience of subjectivity and identity formation. However, twins are *not* doubles. Unlike the figure of the doppelgänger, even so-called 'identical' twins are increasingly used to chart nuanced gradations of difference rather than sameness. Far from a static sameness, the relationship between the twin sisters in *Twinsters* is characterized by a process of recursion. Like the algorithmic systems of content curation and advertising, the image of the self that is reflected back to Samantha in the form of Anaïs provokes a fundamental shift in her life and binds them both into a recursive loop of mutual influence and entwined becoming.

Twinship has also become a major vector for the peculiar mode of stardom cultivated in the digital age. A growing number of social media influencers are using their identity as twins as the cornerstone of their brand. At the time of writing, the Dolan Twins, teenage brothers from New Jersey, currently have 10 million followers each on Instagram, while Texan twins Brooklyn and Bailey have 6 million followers together.[4] A key factor in the visual appeal behind the success of social media twins such as the Dolans and Brooklyn and Bailey is the challenge they pose to their viewers' ability to distinguish between them. The images of twin YouTube stars often intentionally elicit an ambiguity surrounding who is who, or whether or not the siblings have the same tastes and habits. The appeal to the confusion of identities has been one of the mainstays of the twin trope in mass culture from its incarnation in early modern theatre through its use in mistaken identity plots of 1940s Hollywood to the reinvention of these tropes in cinema of the 1990s. All of the identical twin social media stars have posted videos in which they both court uncertainty about their separate identities and lay bare their distinguishing characteristics. The difference from earlier incarnations of twin stardom is that, in the context of social media, the ambiguity surrounding who is who is closely intertwined with an ambiguity surrounding where the human celebrity protagonists end and the technological infrastructure of networked digital media begins.

This confusion is most striking in videos, posted by the Dolan twins and Brooklyn and Bailey, in which both sets of siblings test whether or not they can tell each other apart in old childhood photographs. In Brooklyn and Bailey's vlog post, titled 'Twin Which Twin', the two sisters are shown in what appears to be a private domestic space in front of a pink laptop computer.[5] They react to photos as they appear on their screen and take turns to identify who is who. The photographs also appear in the top left of the screen so that viewers can play along at home. The video contains the main distinguishing hallmarks of teen YouTuber videos. They present themselves as authentic and candidly open their private selves to their followers, inviting viewers into their home and talking them through their family photograph albums. Like most YouTuber videos, Brooklyn and Bailey's vlog post is also highly self-conscious about the active role of different media platforms in shaping individual and collective identities.

Alongside the usual promotion of content distributed through other platforms (in this case, their songs on musical.ly) the video foregrounds its intermediality by reflecting on the competing claims on the real made by different media (in this case, digital vs. analogue photography). Each photograph momentarily throws the separateness of the girls' identities into doubt before the confusion is resolved by the revelation of the correct answer. Brooklyn and Bailey reinforce the identification between stars and viewer by putting themselves in the same position as their followers, struggling to identify who is who. In their doubts and mistakes, they enact the viewer's hesitation before defusing it. The pleasure for viewers here is what Lisa Zunshine describes as the 'cognitively enjoyable' exercise offered by early modern twin plays such as Shakespeare's *Twelfth Night* to have our 'essentialist biases challenged and then confirmed as the distinctions between twin siblings are blurred only to be ultimately reaffirmed'.[6] In the comments to Brooklyn and Bailey's video, followers express the satisfaction of having essentialist biases confirmed when they gloat that 'I can spot Brooklyn better than herself.'

The hesitation evoked by the twin stars in these videos momentarily casts doubt on the border lines between their individual identities. In this, the 'twin challenge' videos are indicative of the ways in which the changing nature of connection in digital cultures is forcing a shift in how we distinguish the individual from the collective, the node from the network. The videos enact a series of slippages: between one face and another, between the technologies of connection and the individuals being connected by these technologies. YouTubers present themselves as the exemplary network labourers of the digital economy: forging connections between disparate t(w)eens behind their screens, brokering alliances between products, identity categories and consumers in endless feedback loops, communicating effortlessly across the various platforms of digital life while evidencing fluency in the always-emerging hybrid languages that each platform demands. In this, YouTube stars are the embodiment of 'nodal citizenship', a term Grant Bollmer has coined to describe those who efficiently carry out the current economic and social imperative to 'relate to others by connecting and maintaining flows' of data.[7] Increasing connectivity in this regime is presented as an innate need of human nature. In the process, the agentic role of technology – the ways in which technologies embody and reinforce ideological discourses – is elided. 'The ability to distinguish between human and technology is eroded, producing humans as objects that serve as imagined material relays supposedly interchangeable with infrastructure.'[8]

Social media are central to this conflation between humans and technology, producing users as 'posthuman' through a 'deeply ingrained and ultimately quotidian belief that it is in human nature to connect and circulate flows of information and capital.'[9] Within this regime, twin YouTubers present themselves as the ideal of posthuman connectivity. The ubiquitous technologies of digital communication in their videos merely reinforce the 'natural' connectedness that exists between them as twins. Rather than an aberration, the twin stars of social media present themselves as taking to an extreme an innate human predisposition towards interconnectedness. Brooklyn and Bailey's ability to answer each other's questions and their simultaneous use of the same hand gestures perform an ideal of frictionless communication, a noiseless transmission of information between emitter and receiver.

Twin YouTubers encourage their followers to think of them simultaneously as one entity and as separate individuals. It has become standard for twin social media stars to have two separate accounts on both Instagram and TikTok, alongside their shared accounts, through which they cultivate personal styles and followings. Many fans express allegiances to one of the twins over the other. The alternation between togetherness and separation that characterizes the rhythms of twin social media branding parallels an alternation between node and network as the stars by turn present themselves as humans with traumas and tastes all of their own and as mere relays in the flows of digital information, interchangeable with the technological infrastructure of social media. Furthermore, twins become a model for an endlessly modulating identity that seamlessly adapts to the shifting categories chosen for it by a network of advertising and curation algorithms. Brooklyn and Bailey adapt to each other's behaviour in a way that echoes the relationship between the user and the online world of consumption of which they are the gatekeepers.

Because of the conceptual challenge they pose to modern individualism, representations of twins often highlight the main configurations of humanity and definitions of what it means to be human in a particular historical moment. In his study of the rituals of the Ndembu people in the Northwest of what is now Zambia, anthropologist Victor Turner argues that twins cause a form of 'embarrassment' since 'there is a classificatory assumption that human beings bear only one child at a time and that there is only one slot for them to occupy in the various groups articulated by kinship which that one child enters by birth'.[10] The 'paradox of twinship' for Turner is that, due to the fact that they are 'hard to fit into the ideal model of the social structure', they often 'become associated with rituals that exhibit the fundamental principles of that structure'.[11] The treatment of twins, he concludes, 'takes on a contrastive character analogous to the relationship between figure and ground in Gestalt psychology'.[12] Turner's work is an example of how anthropologists have often formed twins into what William Viney calls 'a lighthouse group' that, through the treatment of its anomalous nature, is 'used to shed light on others'.[13] Twins have been similarly instrumentalized in literary and cinematic narratives as well as academic studies.

Hillel Schwartz makes this point in his discussion of the role that twins play in what he describes as the 'culture of the copy'. Schwartz argues that two ways of figuring twins are particularly revealing of their respective epochs. In the United States of the mid-nineteenth century, 'during an aggressive period of industrial capitalism', the conjoined 'Siamese Twins' and freakshow stars Chang and Eng Bunker, 'configured by modern notions of the true self as independent and freely willed, evoked the terror of an antithetical, unshakeable double'.[14] Fears of twins, he argues, resurface in the context of the Cold War when twins 'reminded you of doubleness [...] they were antisocial, maybe inhuman'.[15] The end of the twentieth century finds its own version of the twin myth in the legend of the 'vanishing twin', the increasingly common belief that a vast proportion of singletons actually had a twin sibling who died and was incorporated into the body of the surviving twin at an early point of gestation. The vanishing twin myth, Schwartz claims, is an index of a profound crisis in our sense of uniqueness as human individuals. In a society that has become increasingly skilled at replication, vanishing twins simultaneously articulate our profound uneasiness with 'postindustrial contusions of the "real thing"' and compensate for this uneasiness through the assurance of a 'sempiternal human link'.[16] 'At issue,' Schwartz concludes, 'is how each era reconceives, and represents, the naturally human.'[17]

Twins have played a central role in both the construction and undoing of modern subjecthood. On the one hand, they have been used to embody humanity's control and subjugation of nature. Instrumentalized in the twin method of genetics research, dizygotic and monozygotic twins have been conscripted into a scientific discourse aimed at disentangling the influences of culture and nature over human development. Rooted in Francis Galton's experiments of the 1870s, the twin method became a key tool of the eugenics movements of the first half of the twentieth century, a role that was taken to a monstrous extreme in experiments on twins carried out by doctors in the Nazi death camps. Popular culture cemented this association with biopolitical control by establishing a connection between twins and clones. In Aldous Huxley's 1932 dystopia *Brave New World*, for example, the clones are described as 'twins', the splitting of the zygote into two embryos that produces the conditions for monozygotic twins used as an analogy for the technological replication of life.[18] As Anna Tsing reminds us: 'Self-replicating things are models of the kind of nature that technical prowess can control: they are modern'.[19]

On the other hand, twins have also embodied the entanglements disavowed by modernity's positioning of the human subject as separate from and superior to the non-human world. Mythic narratives associate twins with animals, while writers and artists since the Renaissance have exploited the dramatic potential of the challenge posed to human individuality by twins, whether it be early modern comedies of mistaken identity or gothic fiction's monstrous doubles. While twins are a key instrument of modern science's attempts at controlling nature, they have also been an important technology of the non-modern self: the self that is composed in continuous and ongoing relation with the non-human world. It is this construction of twins that has become particularly prominent in our current moment in which the modern myths of individuality and human exceptionalism are being undermined like never before. This attack has taken place on two main fronts: the displacement of individual actors in favour of socio-technical networks in digital cultures, and the entanglement of human and non-human agencies revealed by the current climate crisis. In an era of assemblages, the *entwined being* of twins is replacing the *bounded being* of the individual as the dominant model of subjectivity.

Entwined Being: Twins and Recursion in Digital, Visual and Literary Cultures explores the role of twins in contemporary culture to embody modes of being specific to the cybernetic imaginary of our era of connectivity. While Schwarz traces connections between myths about twins and cultural anxieties about replication and copying, the focus of this book is how twins are used to think through the logic of recursion, one of the defining characteristics of cybernetic systems. Chinese philosopher Yuk Hui argues that 'we are more than ever living in an epoch of cybernetics'.[20] The dominant technological systems of our age – including 'future smart cities, artificial intelligence, machine learning, nanotechnology [and] biotechnology' – all operate through a process of recursion.[21] In his book *Recursivity and Contingency*, Hui makes a distinction between recursion and repetition:

> Recursivity is not mere mechanical repetition; it is characterized by the looping movement of returning to itself in order to determine itself, while every movement is open to contingency, which in turn determines singularity. We can imagine a spiral form, in its every circular movement, which determines its becoming partially from the past circular movements, which still extend their effects as ideas and impressions.[22]

Through this looping movement of recursion, 'the opposition between being and becoming is sublated': 'Being is preserved as a dynamic structure whose operation is open to the incoming of contingency: namely, becoming.'[23]

In a way that echoes Norbert Wiener's account of cybernetic organisms, which is discussed in Chapter 1, Hui uses identical twins as a model for explaining how the play of recursion and contingency constitutes the 'singularity of every living being': 'Why are all twins, in spite of their resemblances, singular on their own?'[24] Hui's metaphor focuses on how twins are set on diverging developmental paths through their differential encounters with contingent events such as 'information [that] triggers the process of individuation'.[25] The manner in which 'identical' twins diverge exposes the true nature of being as a process of becoming. The metaphor does not account for how the individualized becoming of each twin constitutes a contingent event in the development of the other twin and how they are bound to each other in their recursive loops. For many of the texts, films and cultural phenomena discussed in this book, twins provide a model of entwined being that is characterized by recursive entanglements both with each other and their environments.

Biocultural being and molecular recursion

Despite its unshakeable associations with the eugenics movement and repeated attempts to discredit it, twin studies still provide the dominant methodology for parsing genetic and environmental influences on human development. The methodologies associated with twin studies have survived the paradigm shift in the biomedical sciences from a focus on the human body at the 'molar' level to a focus on the 'molecular' body. Whereas, as Nikolas Rose employs the terms, the molar body of nineteenth-century biology was conceptualized in terms of depth – what the body's surface symptoms reveal about its inner workings – contemporary biology operates in a 'flattened' field of open circuits, focusing on 'the functional properties of coding sequences of nucleotide bases and their variations [...] the formation of particular intracellular elements [such as] ion channels, enzyme activities, transporter genes, membrane potentials' that traverse the body's boundaries, disturbing the barrier between its inside and outside.[26] The twin method, as it was developed by Francis Galton during the 1870s, was bound up with a 'truth discourse' grounded in a search for 'hidden entities that determine us'.[27] The premise of this research strategy is to compare the resemblance of monozygotic twins (which result from the division of a single fertilized ovum in early embryonic existence) and dizygotic twins (which arise from the fertilization of two separate ova) on the assumption that the former are genetically identical whereas the latter share the same proportion of genetic material as non-twin siblings. For any particular trait, the greater the difference in concordance between monozygotic and dizygotic twins, the greater the heritability of that trait.

Despite the fact that neither the concept of genes nor the distinction between monozygotic and dizygotic were understood at the time, Galton's 1875 article 'The History of Twins, As a Criterion of the Relative Powers of Nature and Nurture' is both the predecessor of the twin method (which was first fully articulated by the German dermatologist Hermann Siemens in 1923–4) and exemplary of the era's modular approach to the body. In the article, Galton describes an experiment he conducted using twins to determine whether certain characteristics were inherited or obtained through experience and exposure to the environment. Galton sent out surveys to twins or people who were related to twins (he received eighty responses, of which he focuses on thirty-five) asking questions intended to divide their twinship into one of his three categories (strongly alike, moderately alike, extremely dissimilar) while also trying to ascertain whether the twins became either more or less alike as they developed. Based on the results, Galton makes two observations. Firstly, twins that are similar during childhood rarely become less so during maturity, even when they are exposed to different environments. Secondly, twins who were dissimilar at birth never became more similar over time even when they were raised in the same environment. He takes both of these points to conclude that nature has a more significant effect than nurture on human development. This idea, the result of the first recorded twin study, would provide the foundation stone for the eugenics movement. It would also condemn twins to a lasting reputation as a scientific tool for revealing the human body's hidden secrets. Through the Galtonian method, twins seemed to provide a window onto the inheritance buried within us that determines our futures.

Twin studies in the current era of the molecular body have retained this faith in twins as an indispensable epistemological tool. Tim Spector, a professor of genetic epidemiology at King's College London and the director of the largest twin study registry in the UK (TwinsUK Registry) which includes data from 12,000 twins, claims that 'by studying twins, you can learn a great deal about what makes us tick, what makes us different, and particularly the roles of nature versus nurture that you just can't get any other way'.[28] But whereas, in Galton's work, twins seemed to illuminate the determining role of inheritance on human development hidden beneath the surface effects of environmental change, for researchers such as Spector, twinship provides a means of untangling the complexly enmeshed interrelation of genetic and environmental factors. The growing number of multiple births due to the increased availability of IVF treatment is taken as a visible sign of the shift in our conception of what Rose calls 'life itself'. Twinship draws attention to the changes resulting from 'our growing capacities to control, manage, engineer, reshape, and modulate the very vital capacities of human beings as living creatures'.[29]

However, while seeming to embody the perspective on the molecular workings on the body afforded by biotechnology, twinship retains the almost magical promise of revealing the mysteries of 'what makes us tick'. As Viney points out, the endurance of the twin method is due in large part to the fact that, in scientific research, twins are 'made to operate across and between scales', between the molar and the molecular, between scientific truth regimes and 'rather more ancient forms of curiosity, strangeness and exoticism'.[30] While, on the one hand, twins are used to identify processes and developments at the molecular level (at the level of the genome), they also 'constitute the molar and embodied evidence of such research [and are] recruited as rhetorical devices and narrative protagonists by which to prove that those entities, processes, and locations can become publicly understood'.[31]

Due to their status as boundary figures between molar and molecular conceptions of the body, twins cast light on the condition of biocultural being in an age of digital networks. Samantha Frost uses the term 'biocultural creatures' to identify a convergence in how dominant trends within two often distinct genres of research – social theory and the life sciences – are conceptualizing the interaction between the human organism and its environments. On the one hand, a number of interconnected fields of critical theory, from feminist and posthumanist theory to the philosophy of technology, have long sought to displace the anthropocentric view of the human subject as separate from and superior to the natural and technological world in favour of a perspective that privileges constitutive entanglements between humanity and its others, both natural and technological. The non-human turn in the humanities and social sciences has developed in response both to the increasing incorporation of communication devices into everyday life and attempts to understand the implications of anthropogenic climate change for the interrelation between human societies and the natural world (through the concept of the Anthropocene, for example). This non-human turn has drawn from and crystalized the non-anthropocentric tendencies within these various theoretical traditions that have intertwined throughout the twentieth century. On the other hand, a range of fields within the life sciences – including endocrinology, epigenetics and immunology – are finding that 'humans are not simply embedded in their lived environments but that they compose and recompose themselves biologically – at molecular and cellular levels – in response to them.'[32] In other words, the organism and the environment are bound together in a recursive feedback loop.

The field of scholarship emerging from the confluence of these two perspectives conceives of biology 'not as a stable something that subsists doggedly beneath its varied social and political guises' but rather as 'a collection of dynamic processes that are sensitive and responsive to the series of habitats through which humans sustain themselves – and thus as a contributing element in the formation of humans as persons'.[33] In the perspective afforded by this research, rather than the hidden determining force of human development, the genome is viewed as, in Evelyn Fox Keller's words, an 'exquisitely sensitive and reactive system' that develops and mutates in response to 'the constantly changing signals it receives from its environment'.[34] For Frost, the study of human development and the formation of subjectivity must take into account both the formative role of genetic information and the constitutive effect of social interaction upon the biological body (even if the adjustments of the body are not contemporaneous with the environmental causations). So the 'being' of 'biocultural being' is not static but constantly moving and doing, in a perpetual state of composition and decomposition at the dynamic interface between biological and cultural forces.

The role played by twin studies in the wider cultural imaginary of what it means to be human has changed in step with this shift in perspective. Writing in the mid-1990s, Lawrence Wright argued that, in their use in scientific research, 'twins have been used to prove a point, and the point is that we don't become. We are'.[35] The immediate context for Wright's statement, and the focus of much of his book, is the huge amount of publicity surrounding the study of twins reared apart led by Thomas Bouchard in the University of Minnesota. The Minnesota Study of Twins Reared Apart (MISTRA), which ran from 1979 to 2000, gathered information on hundreds of twin siblings who were raised in different households, mainly without knowledge of each other's existence. As well as being its primary object of study, twins who had been reared apart were also a key component of the project's extremely effective publicity machine. One of the first pairs to be studied, in 1979, were James Alan Lewis and James Allen Springer, monozygotic twins separated at four weeks old and only reunited thirty-nine years later. Despite their long separation, the two brothers led extraordinarily similar lives: both named their first son James; both had a dog called Toy and an adopted brother called Larry; both worked in law-enforcement and had an interest in carpentry. The Springer twins feature prominently in nearly every account of the MISTRA project, including in *Born Together, Reared Apart* (2012) by Nancy Segal, a scientist who worked on the Minnesota project and went on to make a career writing popular books about twins.[36] The stories that emerged from the project, such as that of the Springer brothers, seemed to be startling and undeniable proof of the determining force of inherited genetic information, confirming Galton's conclusion of 1875. Although there were plenty of dissenting voices and conflicting studies in the field at the time Wright was conducting his research, the point that twin studies were most often used to prove was that our lives are mostly determined at birth. 'We don't become. We are.'

Twenty years later, in the age of the molecular body, the picture is very different. The focus of twins research is firmly on the complex interplay of biological processes with the environmental factors of culture and social interaction. In the time since Wright's study, twins have come to embody the interface between genetic information and one particular and increasingly dominant facet of social interaction: the production and distribution of digital information. The positioning of twinship at the point of intersection between the genetic and the digital has taken place in both scientific research and popular culture. A number of recent studies in the field of behavioural genetics have used twin studies methodologies to explore genetic predisposition to social media use. In an article titled 'Personalized Media' published in 2017, a team of behavioural genetics researchers based in University College London and the University of Warwick, Ziada Ayorech et al., argue that 'young people choose their online engagements in line with their genetic propensities'.[37] The study draws its data from the Twins Early Development Study, a UK-representative sample of twins born in England and Wales between 1994 and 1996. At the time the research was carried out, more than 10,000 twin pairs remained actively involved in the study. The authors pit their research against what they describe as 'popular media effects theories' that 'typically view the media as an external entity which has some effect (either good or bad) on helpless consumers'.[38] Instead, they claim that their findings 'support an active view of the media environment where individuals tailor their online media use based on their own unique genetic propensities'.[39] They also position their research as pioneering an approach 'that acknowledges media use as a dynamic, human adaptation to the environment where both genes and environment play an integral role'.[40]

The conclusions drawn from the article echo one of Frost's central points: that we are seeing a convergence between social theory and the life sciences in how the interaction between human subjects and their environments is being conceptualized. The idea that we adapt media to suit our needs is one of the central premises of the cultural studies movement in its attempt to distance itself from the Frankfurt School's vision of industrialized mass culture programming a passive public to accept capitalist values. Stuart Hall's work on 'encoding and decoding television discourse' of the 1970s, for instance, emphasizes the reception of mass culture as an active enterprise. In the article on 'Personalised Media' written by Ayorech et al., twins themselves function as a type of medium. Through their entwined being, they become the most effective channel through which to communicate the interrelation between genetic predisposition and the affordances and constraints of digital media.

In a way that foreshadows and mirrors the field of behavioural genetics, in popular mass cultures twins are also increasingly used to map out the emerging points of intersection between genetic information and digital information. Twinship has long been a favoured narrative device of the horror genre, drawing on its associations with the Freudian uncanny and the gothic trope of the double. In a number of recent innovations in the genre – including the films *The Unborn* (2009) and *The Forest* (2016) and Brian Aldiss's 1977 novella 'Brothers of the Head' – it has been used to trace the effects on subjectivity and embodiment of the proliferation of media devices. Through their association with eugenics, cloning and the psionic abilities of telepathy and telekinesis, twins have also proliferated at the fringes of science fiction, performing key conceptual work in the genre's framing of digital culture. Twins constitute a subterranean motif in the work of Philip K. Dick, for example, who, alongside William Gibson, has had the most influence over cultural conceptions of how technologies are reshaping subjectivity and society. Twins also recur in more recent science fiction produced in the Dick tradition, including the 'pre-cog twins' used to divine the future in Steven Spielberg's 2002 adaptation of the short story 'The Minority Report', the twin henchmen in Lily and Lana Wachowski's 2003 sequel *The Matrix Reloaded*, the twin ambassadors in China Miéville's 2011 novel *Embassytown*, and the cloned 'twins' in the Canadian TV series *Orphan Black* (2013–17). Many of the digital twin narratives traced through this book operate on the boundaries between the horror and science fiction genres, using their conventions to explore and shape emerging socio-technical assemblages.

Biopolitics from discipline to recursive modulation

Twin studies have always been one of the key ideological battlegrounds on which the political implications of genetics research are contested. The evolution of the field from the MISTRA project to the present has paralleled the consolidation of neoliberalism. The findings of the Minnesota study, simplified and popularized by figures such as the Springer twins, have been used to support ideas about genetic determinism. In this sense, Wright observes, twin studies played a key role in the 'profound conservative shift' in the United States during the second half of the twentieth century. As the logic goes, 'if people's destinies are written in their genes, why waste money on social programmes'.[41] One of the main financial backers of Bouchard's research was the Pioneer Fund, a New York-based foundation with roots in the eugenics movement of the 1930s and a long track record of advocating racial separation. As the MISTRA project captured the headlines, the most vocal detractors of twin studies – such as Steven Rose, Richard Lewontin, and Leon Kamin, authors of *Not in our Genes: Biology, Ideology, and Human Nature* (1985) – issued warnings about the connection between biological determinism and the political rise of the right. Similarly, a number of projects contemporaneous with Bouchard used twin methods to undermine the determinist argument by complexifying our understanding of the influence of genetic information on human development. This critique has proceeded either by asserting the role of environmental factors or by undermining the conceptual premises of the twin method (such as the simple distinction between monozygotic and dizygotic twins).

Ayorech et al. demonstrate that the twin method maintains a complex relationship with neoliberal ideology. Most obviously, the main thrust of the argument is in support of greater individualization. The future conjured by the article is one in which, thanks to research projects such as theirs, consumers will have the ability to tailor their media devices and software to their own needs, in line with their genetic disposition. Just as online advertising is becoming increasingly personalized, thanks to the production of digital information whenever we shop or browse on the internet, more freely available genetic information will enable hardware and software producers to offer personalized products that cater to our specific needs. As the intersections between digital and genetic information are becoming more and more accessible, they are increasingly becoming objects of interest to advertisers. In September 2018, the music streaming service Spotify announced a collaboration with the commercial DNA testing company AncestryDNA offering to provide playlists tailored to your specific DNA profile. The campaign appeals to notions of genetic determinism while enacting a logic in which our bodies are becoming further enmeshed with the workings of capitalism. As Sarah Zhang put it, marketing campaigns such as this 'trade in the prestige of genomic science, making DNA out to be far more important in our cultural identities than it is, in order to sell more stuff'.[42] While the Spotify-AncestryDNA collaboration is little more than a canny gimmick, it evokes the very real biopolitical implications of the confluence of biotech industries and the network technologies of the information age. The main outcomes of the intersection between biology and informatics that took place through such enterprises as the Human Genome Project, including online genome databases, gene-finding software and genome-sequencing computers, opened up the body to further biopolitical control. Biopolitics, Eugene Thacker reminds us, 'reconfigures the biological domain and "life itself"' through the 'lens of informatics' as something that is 'open to intervention, control, and governance'.[43]

Ken Follet's techno-thriller of 1996, *The Third Twin*, explores the ideological and biopolitical implications of twin studies in the digital age. The novel examines some of the same contexts, characters and concerns as Wright's book, with which it is almost exactly contemporaneous. The plot takes place in a fictitious university in Baltimore and its protagonist Jeannie Ferrami is a post-doctoral scientist using twin studies methods to research genetic predispositions to criminality. Like Wright, Follet emphasizes the politics of twin studies. Ferrami is carrying out her research using data drawn from a big twin studies project, clearly based on MISTRA, run by a private biotech institute called Threeplex whose genetic determinist convictions are intimately intertwined with its conservative politics. While happy to use their funds and resources, Ferrami does not share Threeplex's ideas about the determining influence of genes and instead wants to use her research to prove the decisive role of upbringing in the development of criminal tendencies. 'It's not just your DNA that makes you what you are. It's your upbringing too. That's what my work is all about.'[44] The plot revolves around Ferrami's discovery and exposure of Threeplex's dirty secret: that it used one of its fertility clinics during the 1970s to implant eight cloned embryos into eight unsuspecting women and that these clones are now adults and living out their lives in ignorance of their biotechnological origins. But what Follet picks up on, which Wright does not, is the role of information technologies in opening up the field of genetic research, such as twin studies, to biopolitical manipulation. The main reason Jeannie was offered the job with Threeplex was that during her PhD she developed a computer program that scans medical databases for matching pairs. As she explains:

> It occurred to me that medical information about millions of people is nowadays held on huge databases by insurance companies and government agencies. That includes the kind of data we use to determine whether twins are identical or fraternal: brainwaves, electrocardiograms, and so on. If we could search for pairs of similar electrocardiograms, for example, it would be a way of identifying twins.[45]

While the partners of Threeplex see the potential of Ferrami's software, it is their lack of computer know-how, their failure to understand how information is networked, that proves their downfall. The title of the book performs a slippage between twinship and cloning in an indication of the potential of the former to be instrumentalized in projects of biopolitical control.

Twinship in *The Third Twin* gives expression to a moment of transition between seemingly conflicting power regimes. Twins, in the book, occupy the junctures between the regime of disciplinary power, as characterized by Michel Foucault in *Discipline and Punish*, and the emerging regime of power sketched out by Gilles Deleuze in his 1992 essay 'Postscript on the Societies of Control'. On the one hand, they are subject to imprisonment in the Baltimore prison system. When one of the 'twin' clones is imprisoned overnight, Follet narrates the judicial process in intricate detail. But they are also networked beings. Their bodies traversed by the technologies of medical scrutiny and connected through Ferrami's database-searching software. David Savat identifies what he describes as a 'twin process' at work in the constitution of subjectivity in contemporary cultures of connectivity. 'A desire, and requirement, to constitute oneself as a form – the individual – is combined with and functions alongside a desire, and requirement, to constitute oneself as formless, or to dissolve oneself, to become flow/s.'[46] The 'requirements' of these two processes seem to be antagonistic and incompatible. Whereas one shapes the individual as a subject and object of power, the other works at the sub-individual level, examining patterns of information – from genetic information to consumer data – to anticipate and influence events before they occur. If, in Foucault's account, the diagram of disciplinary power is the panopticon prison, the basic unit of which is the individual criminal, then the diagram of the regime of control is the database containing abstracted information (such as credit ratings and medical information) that can be searched and networked with other databases.

Unlike that of discipline, the subject of the modulatory mode of power is not cellular but 'composed of code'.[47] Digital devices and the media platforms for which they become vehicles often simultaneously work to the logic of both forms of power, producing both modes of subjectivity in parallel. 'With the concurrent operation of discipline and modulation a person is, in one and the same moment, both produced as whole, as a distinct and solid form, while on the other hand is always multiple, always more than one, an infinite dispersal of patterns or lines of code.'[48] To name the subject that is caught between these seemingly contradictory operations Savat uses Gilles Deleuze's term 'dividuality'.[49] Similarly, C. Colwell argues that contemporary culture produces the conditions for 'dialectic without synthesis' or 'hyperdialectic' between individuality and dividuality in which 'the goal is to maintain discursive speed [...] between the two movements'.[50] The system only ceases to function when this incessant shuttling back and forth breaks down and 'leaves us trapped either in individual identity or lost among the dividual fragments of our lives'.[51]

The process of modulation that characterizes control has a recursive structure. John Cheney-Lippold argues that the growing number of data analytics firms, which compile information on individuals based on internet activity and develop computer algorithms to make sense of that data, are producing new forms of 'algorithmic identity'. The constant monitoring of the choices made by internet users and updating of content to fit these choices has created a 'cybernetic relationship to identification'.[52] Increased capacity to aggregate user data has enabled a form of 'real-time cybernetic modelling' in which, 'as more data is received about a certain user's behavior online, new coded computations can be done to change who the user is believed to be and what that user might desire'.[53] In contrast to the forms of stable and enduring subject formation of disciplinary power, the system of control that emerges through a networked society functions through what Cheney-Lippold describes as 'cybernetic characterization' in which 'users are not categorized according to one-off census survey data but through a process of continual interaction with, and modification of, the categories of identity used to target potential consumers'. These categories are not fixed to an ideological ideal imported from the offline world but are constantly modulating in response to new information. And this recursive modulation is a form of control since the futures that are offered to users in the form of advertisements or new content are tethered to their previous patterns of consumption and online activity.

Social media is characterized by the concurrent operation of these antagonistic yet complementary modes of power. Instagram, Facebook and Twitter require their uses to construct themselves as rounded individuals, continually narrating to the world the story of their lives, while at the same time reducing the individual to Bollmer's nodal citizen, a mere relay for digital information. Social media stars are exemplary dividuals. On the one hand, they maintain the cult of the individual through the allure of charismatic celebrity, while, on the other hand, they present themselves as the embodiment of efficient networked capitalism, producing network connections while being reducible to prodigious flows of data. The predominant mode of affectivity in YouTube celebrity culture is that of intensity and preparedness to react to constantly updating information and perpetually changing fashions and trends.[54] This tension in social media stardom can be explained by Tiziana Terranova's account of the role of the individual in the modulatory mode of power. No longer a 'model' for social organization, the individual has become a 'tool that allows the overcoding and the ultimate containment of the productive power of flows'.[55] Individuality and networks are the dual systems that regulate power in the digital age. 'To the decoding of the mass into a network culture, to the dissolution of the individual into the productive powers of a multitude, corresponds an over-coding of the multitude onto the individual element understood as a unit of code'.[56]

The figure of twins has been used to visualize the recursive systems of control described by Cheney-Lippold. The term 'digital twin' has come to refer to virtual simulations of machines or systems that use data capture technology to update in real time. 'Digital twins' allow the operators of these machines or systems to monitor how they are functioning in the present. Furthermore, with the help of machine learning algorithms, they enable them to predict how they will function in the future, how they will react to changes in environmental conditions or other contingencies. The twin in this case becomes a tool of control over the future. However, twinship has also been used to disrupt systems of control. Twin YouTube stars such as the Dolans and Brooklyn and Bailey expose the cracks that exist between the two 'desires' or 'requirements' of these intersecting modes of power. The aesthetic of hesitation they court in their videos – in which the viewer is caught between attempting to assert their separate identities and regarding them as one entity united by their brand and, consequently, at one with the flows of capital and digital data – turns the alternation between discipline and modulation into a cognitively enjoyable exercise, to use Zunshine's phrase. While, ultimately, social media stardom serves to naturalize the state of dividuality, the hesitation produced by the twin stars opens up a space for its critique. These social media twins produce what Savat describes as a moment of 'unharmony' in the coexistence between these conflicting 'desires' or 'requirements', a tremor that exposes the emergence of new forms of subjectivity or assemblages at the interface between bodies and machines.

The figure of twins in the age of cybernetics

The function of the figure of twins in the age of cybernetics echoes that of the sibling during the nineteenth century. In *Sibling Action*, Stefani Engelstein argues that the category of the sibling is indicative of both the identity politics and knowledge systems of modernity. During the long nineteenth century (from about 1770 to 1915), the figure of the sibling served as both 'an epistemological tool for differentiating terms' and 'an educational tool for directing affect and desire'.[57] On the one hand, the genealogical systems into which the sibling category was conscripted had the dual function of 'controlling the contours of kinship' and 'naturalizing – and hence legitimizing – systems of classification and knowledge' grounded in a fundamental separation between self and other.[58] On the other hand, as a tool for channelling desire in ways that support broader systems of capitalism and nationhood, 'sibling affection grounds more generalizable sympathetic attachments and guides ethics'.[59] The ideals of 'fraternity' or 'universal brotherhood' serve to 'balance the opposing forces' that are constitutive of liberal democracy: the needs of free and 'self-interested individuals' with the 'rhetoric of equality'.[60] However, while the sibling is an 'enabler of classification' it is also a 'splinter of instability' that 'serves as a disruptive action within the very system that it underwrites'.[61] For the sibling is a distinctly ambiguous category that is at once part of the self and 'not-quite-other'.[62] And, for that reason, 'it marks the contingency and permeability of boundaries, the doubtfulness of integrity, and the insecurity of uniqueness'.[63]

As opposed to the Freudian system of 'vertical kinship models' that came to dominate the twentieth century, sibling logic 'recognizes the subject as embedded in a transsubjective network of *partial others*': 'The sibling as a model [...] allows us to move beyond both self-other dualisms and the mother-child dyad as the only grounds for intersubjectivity and positions the subject as instead embedded in a network of partial others, whose subjectivities are nonetheless partially, though differentially, shared'.[64] Engelstein reads the networks of siblings in Sophocles's play *Antigone* as a model for what she calls 'synecdochal intertwining' that 'extends subjectivity beyond the immediate family and into the polis and the political'.[65] Building on a textual reading of the play by Simon Goldhill, she argues that the two sisters (Antigone and Ismene) and two brothers (Polynices and Eteocles) are described as 'dual units' that are not, however, 'closed to further linkage' but rather seem to double each other in a series of proliferating connections that extend beyond the family. In the texts and images analysed by Engelstein, amid the uses of the figure of the sibling to police categories and distinctions, there is evidence of a more speculative emphasis on the ambiguity of siblinghood which exposes the 'contingency of classifications' and attempts to move beyond normative kinship structures.[66]

The figure of twins and the logic of twinning traced in this book revive this speculative use of the ambiguity of the sibling, which was largely subsumed during the twentieth century by the Freudian focus on the mother-child dyad. By embodying the entanglements disavowed by individualism, twins have played a prominent role in disrupting the dominant kinship systems of modernity. Viney, for example, points out that the dramatic increase in twin birth rates in the UK and the United States is an index of the growing availability of IVF treatments and other kinds of assisted reproductive techniques. 'Twin births have increased because hormone stimulants promote the production of many eggs, and this can lead to twin pregnancies even before IVF has taken place. Multiple embryo transfer – or transferring two or more embryos back into the uterus – significantly increases the chances of conceiving twins.'[67] Viney takes the example of the three children born through IVF surrogacy to British couple Barrie and Tony Drewitt-Barlow. The eldest two, Aspen and Saffron, are gestational twins in that they shared a womb but are only genetic half-siblings since they had different biological fathers. Though 4 years their junior, Orland is Aspen's embryological and genetic albeit non-gestational twin. In an era of what Viney calls 'time-lapsed IVF twinning', twin identities 'make for increasingly plural relationships and enlarge the ways that people can feel twinned with others, and untwinned from their closest kin'.[68]

Just as the sibling, for Engelstein, opens up affective ties beyond the family, the logic of twinning points beyond kinship. The practice of 'twinning' in networked image cultures, for instance, is characterized not by a desire for sameness but by a movement of recursion through which the subject repeatedly enfolds contingency into its paths of becoming. An example of this is the increasing number of twinning apps that trawl online image databases to find the user's online doppelgänger. The website 'Twin Strangers', for instance, promises to 'find your lookalike from anywhere in the world'. But while the site's promotional images emphasize the 'sameness' of the lookalikes – the fact that they could pass for 'identical' twins – accounts of using the software discussed in Chapter 6 describe a form of similitude that is undercut by fine gradations of difference between the two website users that are amplified and mutate through their recursive interaction with each other. The coherence of the self is undermined through fractal involutions between self and other and this effect is strengthened by the fact that one Twin Strangers client can find multiple lookalikes and, therefore, set off a chain of proliferating doubles. Embodying a tension between sameness and difference at the heart of algorithmic image cultures, the form of twinning proposed by Twin Strangers is a counter-tendency to what Wendy Chun describes as the dominant digital network logic of 'homophily'. This is a term Chun uses to describe how advertising and curation algorithms recursively 'feedback' choices and preferences to users of internet systems and in the process consolidate consumers into 'neighbourhoods' occupied by supposedly near-identical types.[69] Rather than just appeal to pre-existing social types and the divisions between them, these algorithms, Chun claims, are actively bringing them into existence. Although it evokes sameness, twinning carries out what Chun describes as a 'queering' of the logic of homophily by performing a recursive becoming at the heart of internet identity production. In this, then, twinning functions as a guide to revisit the notion of 'kin' and make it 'mean something other/more than entities tied by ancestry or genealogy'.[70]

Furthermore, although the figure of twins has often been conscripted into dualist thinking, through the association with cybernetics traced in this book they also point beyond dualisms. Twins may seem to be inextricable from dualities of various kinds. The trope of the good and evil twin is ubiquitous in popular culture, a usage that Juliana de Nooy and Bronwyn Statham point out is strongly gendered, with 'sisters always split along the same predictable line – a version of the virgin/ whore dichotomy'.[71] By positioning twinning within a logic of recursion, the texts, images and discourses discussed in this book enrol twins in an overcoming of dualism. As discussed in Chapter 2, Eduardo Viveiros de Castro identifies twin relations with a form of 'dynamic instability' or 'perpetual disequilibrium' that undoes dualistic thinking. And yet, twins do not always become figureheads for a celebration of constitutive relationality and an overcoming of dualism. Rather, they become guides to the dangers of cybernetic systems. Hui argues that 'maybe it is no longer a dualism which is the source of danger in our epoch, but rather a non-dualistic totalizing power present in modern technology, which ironically resonates with the anti-dualistic ideology.'[72] While for Engelstein the ambiguous figure of the sibling points to the 'contingency of classification', twins in the digital age occupy the fault lines between the dualistic thinking of modernity and a totalizing machinic organicism.

Overview and structure

Various critics working within and across different disciplines have turned to twinning not just as an object of study but also as a form of engaged methodology. In his study of twins, Viney argues that interdisciplinary scholarship takes shape in relation to 'restless objects of study': 'Some objects of research, like the things we use to make thought more material, cannot and will not stay indoors'.[73] Studying twins is a methodological challenge since their cultural representations have always been so closely intertwined with scientific discourse. Their use as monstrous limit cases of the human in everything from children's literature to gothic horror both echoes and informs their role as objects of scrutiny in genetics research. The laboratory and the freak show are symbiotically connected. As Karen Dillon puts it, 'The narrative of twins as exotic others established in the late nineteenth century converges with the contemporaneous shift in twins' status from biological curiosities to popular research subjects for scientists to debate genetic versus environmental determinism.'[74] Schwartz also points out that advertising has long drawn on the reputation of twins as 'truth-tellers about human nature' resulting from their construction in scientific research as providing a window onto the determining role of our genetic material.[75] Twins in this case connect at least three distinct systems of knowledge: mythological narratives of twins as oracles; a scientific approach centred on the search for correlations and divergences (between monozygotic and fraternal twins); and the commercial language of advertising. The twins that populate this book find themselves entangled within a broad range of discourses, including postgenomic science, algorithmic systems, literature and visual culture. As a result, I have had to draw on an equally broad range of methodologies, predominantly from the fields of media studies and cultural studies, while also engaging with debates within the medical and environmental humanities.

The frequent enrolment of twins in metaphor has also increased their entanglements with multiple fields of knowledge. In their metaphorical incarnations, twins have been made to embody acts of connection (such as the practice of town twinning that was used to foster international cooperation in the wake of the Second World War) and dualism (as in the common use of the twins to refer to two connected ideas or phenomena). Part of their appeal to geneticists is their ability to provide a bridge across discursive domains. They are 'natural experiments' or 'living proof' of the agency of our genetic code. Popular fascination with uncanny similarities between identical twins such as the Springer brothers turned them into free advertising for the MISTRA project. Many of the artworks and novels discussed here tap into the metaphorical potential of twins. And to study the figure of twins is to study how metaphor shapes our world, which is caught, like twins, between dualism and entanglement. But more often than not, twins, such as the figure of glitch twins discussed in Chapters 5 and 6 and the Conclusion, are used to undermine the process of metaphor or expose disconnections and incompatibilities between systems.

Florence Chiew and Alison Barnwell argue that the figure of twins, due to its 'unique troubling of oppositional categories', confronts the researcher with the challenge of what they call 'methodological intimacies', the acknowledgement that 'methods of knowing are always deeply entangled with their objects of investigation'.[76] Through this recognition, the 'prism of twinning' makes the process of research 'something more entangled and unsure'.[77] It is easy to spot this methodological intimacy in the work of others. In the final pages of *The Story of Lynx*, discussed in Chapter 3, the 'dynamic disequilibrium' that he sees in twins induces in Claude Lévi-Strauss a form of hermeneutic delirium or fever of interpretation in which he sees everything as being connected. Tracing connections between associations surrounding twins in both the 'New' and 'Old Worlds' enables him to 'catch a glimpse' of a common 'moral world' though, he admits, 'this is probably only an illusion'.[78] Twins induce a search for correlation. As a father of twins, the entanglement between my scholarly approach and object of study is particularly 'intimate'. And while sounding out these associations between twins and cybernetic thought I have, at times, partially given myself over to the fever of interpretation evident in the final pages of *The Story of Lynx*. While I have restricted my study to the era of cybernetics, from the 1940s to the present, I have traced the influence, resonance and mirrorings of cybernetic thought and the ways in which this has been performed or framed in representations of twins across geographical and linguistic borders. While twins have been used as a metaphor to think transnationally since the Second World War, in the form of 'town twinning', the twins discussed in this book compel us to confront a transnationalism that is much weirder and complex than that envisioned by post-war bureaucrats, the distances and frontiers of which are being radically reshaped by digital connectivity and the global climate crisis. As I will show in this book, twins both seem to point to the pervasive connections in the fabric of this system, and gesture to the limitations of connections, the epistemic blockages and intra-systemic glitches.

The first chapter of this study explores the use of the figure of twins and twinning in cybernetics. I start by exploring how Norbert Wiener and Grey Walter, pioneers in the field of cybernetics, used twins and their relationships with each other as a metaphor through which to compare how human and machine organisms interact with their environments. I go on to examine the trope of telepathic twins in science fiction from the 1940s to the 1980s, including in the work of A.E. van Vogt, Robert Heinlein, Theodore Sturgeon and Philip K. Dick. Against a gothic tendency to portray twins as inward-looking closed systems, characterized by incest and degeneration, these texts describe their telepathic twin characters as open systems caught up, through a recursive looping dynamic, in a constant intra-action with their environments. In the process, they explore the limits of cybernetic control systems. In Chapter 2, the discussion turns to the use, in Weird Fiction, of twins and the bond that exists between them as a form of mediation that blurs the distinctions between the technical and the organic. Through their association with the trope of the doppelgänger, in gothic fiction twinship is often characterized as a channel of communication so uncannily powerful that it can traverse the barrier between life and death. Similarly, in genetics twins are constructed as a form of 'haunted media' that connects human development and behaviour in the present with the agential role of genetic inheritance. In the texts that are the focus of this chapter – which include H.P. Lovecraft's 'The Dunwich Horror' (1929), Jeff Vandermeer's Southern Reach trilogy (2014), the 2009 horror film *The Unborn*, Fumiko Enchi's 1958 novel *Masks* and Peter Greenaway's 1985 art film *A Zed & Two Noughts* – twinship constitutes a form of weird or blocked mediation. In the process, they divert twins from their fate as instruments in the scientific control of nature and open them up to weird transversal connections between humans and their non-human environments.

Chapter 3 explores the use of twins in anthropological theory and literary fiction as guides to the survival of the Anthropocene. In *Cannibal Metaphysics*, Eduardo Viveiros de Castro argues that in Amerindian cosmologies, twins are used as a model not of sameness or symmetry but of the 'dynamic difference' and 'perpetual disequilibrium' that are the driving motors of life. I examine Viveiros de Castro's development of this construction of twinship in dialogue with Claude Lévi-Strauss as well as the concept of 'disjunctive synthesis' in the work of Gilles Deleuze, while considering whether it is both an instrumentalization of twins and a colonizing gesture that strips indigenous thought of its history. I end the chapter by arguing that twins have been used to articulate what Timothy Morton describes as the 'weird temporalities' of the Anthropocene. Shelley Jackson's 2006 novel *Half Life*, the main example I explore in the chapter, uses a narrative recounted by conjoined twins to perform a series of feedback loops between the human body and the environment, which cut through and across different scales of space and time. The role of recursion in these narratives points to the influence of cybernetics in the models of post-anthropocentric subjectivities posited by contemporary critical theory. The focus of Chapter 4 is the role of twinning in the aesthetics of black Futurism. Taking up the theme of weird mediation, I start by exploring how twins, in Chinua Achebe's 1958 novel *Things Fall Apart*, embody a point of incompatibility between conflicting epistemologies. Contemporary Afrofuturist narrative and aesthetics reinterpret this function of twinship as a glitch that emerges through interactions between racial and technical systems. Through close readings of Helen Oyeyemi's 2005 novel *The Icarus Girl*, Tade Thompson's Wormwood Trilogy (2016-19) and the performances of French dance duo Les Twins, I argue that the recursive processes of twinning loop together traumatic histories of slavery with new speculative connections between blackness and digital technologies.

Chapter 6 examines how twins function as glitches within biometric technologies and particularly facial recognition systems. Through a close reading of the performance of twinship as a form of mask by social media celebrities,

I argue that twinning introduces moments of uncertainty into the datafication of identity. This uncertainty is echoed in the practice of twinning in apps, such as 'Twin Strangers', that trawl online image databases to 'find your lookalike from anywhere in the world'. Together with the phenomenon of the twin selfie, these uses of twins undermine the dominant social media logic of what Wendy Chun terms 'homophily' which reinforces social divisions through a consolidation of sameness and homogeny. Unlike homophily, twinning here is characterized by a productive disequilibrium that is the product of recursive interactions between self and digital networks. The Conclusion explores the use of the metaphor of 'digital twins' in the control of complex systems through the prediction of future uncertainties. By contrast, I argue, the twins discussed in this book introduce glitches into cybernetic systems. Rather than the predictable future envisaged through digital twins, these 'glitch twins' produce a looping temporality that disrupts the agenda of control while introducing uncertainty into attempts to classify and separate.

Notes

1 Alissa Wilkinson, 'What's with All the Movies about Doppelgängers?', *The Atlantic*, 14 March 2014, https://www.theatlantic.com/entertainment/archive/2014/03/whats-with-all-the-movies-about-doppelg-ngers/284413/.
2 Wilkinson, 'What's with All the Movies about Doppelgängers?'.
3 Wilkinson, 'What's with All the Movies about Doppelgängers?'.
4 This is the number of followers each had on 24 June 2021.
5 Brooklyn and Bailey, 'Twin Which Twin: Guess the Baby Pictures?', YouTube, 22 February 2017, https://www.youtube.com/watch?v=ppi-c81j7jM.
6 Lisa Zunshine, *Strange Concepts and the Stories They Make Possible* (Baltimore: The Johns Hopkins University Press, 2008), 35.
7 Grant Bollmer, *Inhuman Networks: Social Media and the Archaeology of Connection* (London: Bloomsbury Academic, 2016), 7.
8 Bollmer, *Inhuman Networks*, 8.
9 Bollmer, *Inhuman Networks*, 5.
10 Victor Turner, *The Ritual Process: Structure and Anti-Structure* (New York: Aldine de Gruyter, 1969), 45.
11 Turner, *The Ritual Process*, 80.
12 Turner, *The Ritual Process*, 80.
13 William Viney, *Twins: Superstitions and Marvels, Fantasies and Experiments* (London: Reaktion Books, 2021), 85.

14 Hillel Schwartz, *The Culture of the Copy: Striking Likenesses, Unseasonable Fascimiles* (New York: Zone Books, 1996), 58 and 59. Dror Wahrman argues that attitudes towards conjoined twins index changing conceptions of individuality in the eighteenth century. In her analysis of the *Memoirs of Martinus Scriblerus*, written by John Arbuthnot, Jonathan Swift, John Gay and Alexander Pope, the hero of which falls in love with Siamese twins joined at the sex organ, Wahrman identifies a shift from a preoccupation with the sameness that exists between twins to an emphasis on difference and 'disharmony': 'Was this another manifestation, or projection, of the eighteenth-century interest in the splitting and mutating of identity?' Dror Wahrman, *The Making of the Modern Self: Identity and Culture in Eighteenth-Century England* (New Haven and London: Yale University Press, 2004), 194.

15 Schwartz, *The Culture of the Copy*, 31.
16 Schwartz, *The Culture of the Copy*, 41.
17 Schwartz, *The Culture of the Copy*, 66.
18 Aldous Huxley, *Brave New World* (London: Random House, 2008 [1932]), 4.
19 Anna Lowenhaupt Tsing, *The Mushroom at the End of the World: On the Possibility of Life in Capitalist Ruins* (Princeton and Oxford: Princeton University Press, 2015), 140.
20 Yuk Hui, *Recursivity and Contingency* (London and New York: Rowman & Littlefield, 2019), 278.
21 Hui, *Recursivity and Contingency*, 278.
22 Hui, *Recursivity and Contingency*, 17–18.
23 Hui, *Recursivity and Contingency*, 18.
24 Hui, *Recursivity and Contingency*, 19.
25 Hui, *Recursivity and Contingency*, 19.
26 Nikolas Rose, *The Politics of Life Itself: Biomedicine, Power, and Subjectivity in the Twenty-First Century* (Princeton and Oxford: Princeton University Press, 2007), 12.
27 Rose, *The Politics of Life Itself*, 15.
28 Tim Spector quoted in Ker Than, 'A Brief History of Twin Studies', *Smithsonian Magazine*, 4 March 2016, https://www.smithsonianmag.com/science-nature/brief-history-twin-studies-180958281/.
29 Rose, *The Politics of Life Itself*, 3.
30 William Viney, 'Getting the Measure of Twins', in *The Edinburgh Companion to the Critical Medical Humanities*, ed. Anne Whitehead, Angela Woods, Sarah Atkinson, Jane Macnaughton and Jennifer Richards (Edinburgh: Edinburgh University Press, 2016), 104 and 119.
31 William Viney, 'Experimenting in the Biosocial: The Strange Case of Twin Research', in *The Palgrave Handbook of Biology and Society*, ed. Maurizio Meloni, John Cromby, Des Fitzgerald and Stephanie Lloyd (New York: Palgrave, 2018), 160.

32 Samantha Frost, 'Ten Theses on the Subject of Biology and Politics: Conceptual, Methodological, and Biopolitical Considerations', in *The Palgrave Handbook of Biology and Society*, ed. Maurizio Meloni, John Cromby, Des Fitzgerald and Stephanie Lloyd (New York: Palgrave, 2018), 897.
33 Frost, 'Ten Theses on the Subject of Biology and Politics', 900.
34 Evelyn Fox Keller, 'The Postgenomic Genome', in *Postgenomics: Perspectives on Biology after the Genome*, ed. Sarah Richardson and Hallam Stevens (Durham, NC: Duke University Press, 2015), 10 and 25.
35 Lawrence Wright, *Twins: Genes, Environment and the Mystery of Identity* (London: Weidenfeld & Nicolson, 1997), 9.
36 Nancy Segal, *Born Together – Reared Apart* (Cambridge, MA: Harvard University Press, 2012).
37 Ziada Ayorech, Sophie von Stumm, Claire Haworth, Oliver Davis, Robert Plomin, 'Personalized Media: A Genetically Informative Investigation of Individual Differences in Online Media Use', *PLoS ONE* 12:1 (2017): 2.
38 Ayorech et al., 'Personalized Media', 7.
39 Ayorech et al., 'Personalized Media', 2.
40 Ayorech et al., 'Personalized Media', 7.
41 Wright, *Twins*, 9.
42 Sarah Zhang, 'Your DNA Is Not Your Culture', *The Atlantic*, 25 September 2018, https://www.theatlantic.com/science/archive/2018/09/your-dna-is-not-your-culture/571150/.
43 Eugene Thacker, *The Global Genome: Biotechnology, Politics, and Culture* (Cambridge, MA and London: The MIT Press, 2005), 29.
44 Ken Follet, *The Third Twin* (London: Pan Macmillan, 1996), 19.
45 Follet, *The Third Twin*, 103.
46 David Savat, *Uncoding the Digital: Technology, Subjectivity and Action in the Control Society* (New York: Palgrave MacMillan, 2013), 134.
47 Savat, *Uncoding the Digital*, 34.
48 Savat, *Uncoding the Digital*, 54.
49 In many ways the condition of dividuality captures the experience of being subjected to power in biopolitical regimes. In the Collège de France lectures of 1976, Foucault makes a distinction between discipline and biopolitics: 'So after a first seizure of power over the body in an individualizing mode, we have a second seizure of power that is not individualizing, but, if you like, massifying, that is directed not to man-as-body but to man-as-species.' Rather than one 'seizure of power' succeeding another, the two work in close coordination. Michel Foucault, *'Society Must Be Defended': Lectures at the Collège de France, 1975–6*, trans. David Macey (New York: Picador, 2003 [1976]), 243.
50 C. Colwell, 'Discipline and Control: Butler and Deleuze on Individuality and Dividuality', *Philosophy Today* 40:1 (1996): 216.

51 Colwell, 'Discipline and Control', 216.
52 John Cheney-Lippold, 'A New Algorithmic Identity: Soft Biopolitics and the Modulation of Control', *Theory, Culture & Society* 28:6 (2011): 167.
53 Cheney-Lippold, 'A New Algorithmic Identity', 167.
54 The requirements of intensity and preparedness also account for the most prevalent genre of YouTube stardom: the 'prank'. The inhabitants of the worlds that these stars construct for their viewers must exist in a state of constant alertness, lest some humiliating prank befall them. In the videos produced by 'professional weirdo' Jackson O'Doherty, if he or any of his sidekicks drop their guard for one minute, they will have their trousers pulled down or – the most terrible of all punishments – have their phones thrown away. In this world of pranks rifts are continually appearing in the fabric of reality and those who occupy this world have to rapidly adjust, keeping constantly on their toes.
55 Tiziana Terranova, *Network Culture: Politics for the Information Age* (London: Pluto Press, 2004), 123.
56 Terranova, *Network Culture*, 123.
57 Stefani Engelstein, *Sibling Action: The Genealogical Structure of Modernity* (New York: Colombia University Press, 2017), 7.
58 Engelstein, *Sibling Action*, 22.
59 Engelstein, *Sibling Action*, 229.
60 Engelstein, *Sibling Action*, 68.
61 Engelstein, *Sibling Action*, 27.
62 Engelstein, *Sibling Action*, 3.
63 Engelstein, *Sibling Action*, 3.
64 Engelstein, *Sibling Action*, 38.
65 Engelstein, *Sibling Action*, 51.
66 Engelstein, *Sibling Action*, 9.
67 Viney, *Twins*, 124.
68 Viney, *Twins*, 139 and 141.
69 Wendy Hui Kyong Chun, 'Queering Homophily', in *Pattern Discrimination*, ed. Clemens Apprich, Wendy Hui Kyong Chun, Florian Cramer and Hito Steyerl (Minneapolis and London: University of Minnesota Press, 2018), 75.
70 Donna J. Haraway, *Staying with the Trouble: Making Kin in the Chthulucene* (Durham and London: Duke University Press, 2016), 102–3.
71 Juliana De Nooy and Bronwyn Statham, 'Telling the Good from the Bad in Twin Films', *Journal of Media & Cultural Studies* 12:3 (1998): 280.
72 Yuk Hui, 'Machine and Ecology', *Angelaki* 25:4 (2020): 58.
73 Viney, *Twins*, 17.
74 Karen Dillon, *The Spectacle of Twins in American Literature and Popular Culture* (Jefferson, NC: McFarland & Company, Inc., 2018), 8.

75 Schwartz, *The Culture of the Copy*, 32.
76 Florence Chiew and Alison Barnwell, 'Methodological Intimacies and the Figure of the Twins', *The Sociological Review Monographs* 67:2 (2019): 469.
77 Chiew and Barnwell, 'Methodological Intimacies and the Figure of the Twins', 477.
78 Claude Lévi-Strauss, *The Story of Lynx*, trans. Catherine Tihanyi (Chicago and London: The University of Chicago Press, 1995), 241 and 242.

1

Telepathic twins and the mythotechnesis of cybernetics

The conjoined twin protagonists of Vladimir Nabokov's 1958 short story 'Scenes from the Life of a Double Monster' possess the ability to communicate wordlessly. 'Essential sensations' flow between them like 'shed leaves riding the stream of our shared blood', while '[t]hin thoughts also manage to slip through and travel between us'.[1] The 'bridge' of flesh that connects their bodies has brought with it a world of not only entangled sensation (described as a 'common and continuous rhythm') and unconscious urges ('we sometimes pooled our minds when we dreamed') but also conscious thought and decision-making.[2] As the narrator Floyd puts it: 'quite often problems puzzling me were twofold: possibly a trickle of Lloyd's cerebration penetrated my mind and one of the two linked problems was his.'[3] The figure of the double recurs throughout Nabokov's fiction, from the parodic treatment of gothic doppelgänger narratives in *Despair* (1934) to the use of 'Siamese twins' as a metaphor for the 'anatomically conjoined communities' of democratic nations in *Pnin* (1957). In 'Scenes from the Life of a Double Monster', the trope of telepathic twins is deployed to interrogate the paradoxes encoded within a term that splices distance ('tele') with empathetic connection or touch ('pathos'). For Nabokov, it is Floyd and Lloyd's monstrous proximity that stands in the way of their intimacy by undermining their separateness as distinct entities.

The attribution of telepathic abilities to twins was nothing new when Nabokov was writing his story in the 1950s. Alexandre Dumas's 1844 novella 'The Corsican Brothers' tells the story of conjoined twins who, though separated at birth, continue to communicate telepathically, indicating that, by then, the association was already commonplace. Towards the end of the nineteenth century, communication between twins was frequently cited as proof of the existence of telepathy by the Society for Psychical Research (SPR) who coined the term and defined it as 'impression received at a distance without the operation of the recognised sense organs'.[4] The writings of Frederic Myers, one of the founding members of the SPR, used the concept of telepathy to forge a conceptual synthesis between a range of distinct and emerging epistemological practices and systems, from mesmerism and mediumship to William Crooke's notion of a 'psychic force'. In a text published in 1903 that summarized research conducted since the late 1880s, Myers cites a number of twin cases while laying out his proof for the existence of telepathy. Admitting to a lack of thorough research into the subject, he nevertheless observes that 'one clue (a vague one at yet) to the causes directing and determining telepathic communications lies in what seems their exceptional frequency between *twins*; – the closest of all relations'.[5]

Myers is not alone in citing the unusual 'closeness' of twins as a reason behind their propensity for telepathic communication. In his overview of scientific research into the connection between twins and telepathy, Guy Lyon Playfair, for instance, argues that the 'twin connection' displays 'the telepathic signal at full volume, as it were, at which not only information is transmitted at a distance but so are emotions, physical sensations and even symptoms such as burns and bruises'.[6] For Myers and Playfair, then, telepathic twins are not monstrous aberrations of nature. Rather, they reveal the potential for extra-sensorial communication that is latent within all human minds. It is not a stretch to argue that twins, for Myers, were the embodiment of his theory of 'subliminal consciousness' developed through his research for the SPR. In an essay published in 1892, he argues that: '[e]ach of us is in reality an abiding psychical entity far more extensive than he [sic] knows – an individuality which can never express itself completely through any corporeal manifestation. The Self manifests through the organism; but there is always some part of the Self unmanifested'.[7] Telepathy, then, takes place when tapping into this 'Self unmanifested'. While in 'Scenes from the Life of a Double Monster' Nabokov presents this non-coincidence between physical organism and psychical self as monstrous, Myers holds it up as a marker of evolutionary advance, what he called 'preversion'. However, in both cases, the trope of telepathic twins is used to dismantle the boundedness of the individual self.

Read closely, however, the description of telepathic twins in 'Scenes from the Life of a Double Monster' displays the influence of a set of ideas and technological systems that were emerging into public consciousness during the 1950s under the term 'cybernetics'. In the first description of the entwined being shared by the brothers, the narrator Floyd tells the reader that '[t]he pattern of acts prompted by this or that mutual urge formed a kind of grey, evenly woven, generalized background against which the discrete impulse, his or mine, followed a brighter and sharper course; but (guided as it were by the warp of the background pattern) it never went athwart the common weave of the other twin's whim'.[8] The individual thoughts, decisions and acts emerge out of a 'background pattern' that is common to both Floyd and Lloyd. The monstrous fate that Floyd is trying to escape by narrating the tale is the subsumption of his individual self into this common pattern. The story was intended to be the first part of a novel that ended with Floyd finally being separated from Lloyd and gaining his independence (only for both brothers to die); the urge for separation is what guides the narrative.[9] The choice of words evokes the fate of the individual in the cybernetic account of the continuity between human and non-human informational systems. As Katherine Hayles puts it in her account of the 'posthuman' vision that emerged from the cybernetic theorists of the second half of the twentieth century, 'the posthuman view privileges informational pattern over material instantiation, so that embodiment in a biological substrate is seen as an accident of history rather than an inevitability of life'.[10] The term 'pattern' is key in Norbert Wiener's popular writings on cybernetics of the post-war period. In his articulation of the human self in terms of information (which will be discussed more fully later in the chapter) he states that '[w]e are not stuff that abides, but patterns that perpetuate themselves'.[11] Before Floyd asserts his own individuality, he describes the conjoined bodies of the twins in terms that privilege informational pattern over the existence of individual spirit or soul. In a way that echoes humanist critics of cybernetics, as well as Wiener's fears of the philosophical implications of his own research, Floyd quickly moves away from this vision of human life in terms of 'pattern'.

A striking tension runs through Nabokov's story, which echoes the humanist hesitations evident in the work of early cyberneticists. On the one hand, Floyd and Lloyd's conjoined state is described in terms of control and restriction. Floyd's narration frames their joint existence in terms of a power struggle, both with each other and with the grandfather and uncle who exploit their physical anomalies in a travelling freakshow. The shared corporeal system regulates the actions of the twins. The brothers can give expression to 'discrete impulses' only so far as they do not contravene 'the common weave of the other twin's whim'. This account of control echoes that which was developed in the Macy conferences of the post-war period in which researchers and theorists discussed the wider social implications of cybernetic thought. As Ute Holl explains it, 'Similar to navigating on the ocean – kybernetes, the helmsman, is the Greek etymological godfather of all cyberneticists – understanding behaviour was thought to be a process of orientation in ongoing evaluations and course corrections in complex and interpersonal situations. To do so, a method to discover patterns of behaviour beyond individual actions, perceptions, and teleologies was necessary.'[12] The 'rhythmic' structure shared by the brothers is a structure of systemic control which determines each individual twin's behaviour and without which they cannot exist. On the other hand, Floyd and Lloyd's conjoined body is described in terms of an openness that unbinds the strictures of the individual self. Much of the language and imagery used to describe their shared system evokes that of cybernetic ecological theory. Their shared bodily system mirrors and is bound up with the natural ecologies around them. Individual sensations are 'shed leaves' riding on a common stream; 'discrete impulses' are 'ripples' or 'trickles' that disrupt a shared pool of experience. The critical posthumanist potential of their conjoined state is evoked before being shut down by the narrator's focus on control. The emergence of 'discrete impulses' out of a 'background pattern' foreshadows Félix Guattari's assertion, in *The Three Ecologies*, that '[r]ather than speak of the "subject"', we should perhaps refer to '[v]ectors of subjectification [that] do not necessarily pass through the individual'.[13] The brothers' shared unconscious accessed by common dreams reads like the desired end product of Guattarian schizoanalysis, a radical process of dis- and re-assembling the self.

The ambiguity discernible in 'Scenes from the Life of a Double Monster' is indicative of how the trope of twin telepathy, as it develops from the post-war period to the present, is employed to trace the unstable intersection between cybernetic control and ecological openness. Erich Hörl traces the roots of the ecological turn in critical thought to the development of cybernetics from the post-war period. What he calls 'the ecologization of thinking', and its 'radical revaluation of relationality', emerged from and, to an extent, 'is still inscribed in [...] the history of control and the corresponding rationality of power'.[14] Hörl argues that this historical entanglement between relationality and control has never been more visible than in the current phase of informational capitalism. 'This phase witnesses the emergence of an environmental culture of control that, thanks to the radical environmental distribution of agency by environmental media technologies, ranging from sensorial to algorithmic environments, from bio- to nano- and geotechnologies, renders environmentality visible and prioritizes it like never before.'[15] This culture of environmental control brings with it a 'reduction of relations to calculable, rationalizable, exploitable' metrics.[16] Science fiction narratives about telepathy produced in the second half of the twentieth century constitute an important archive of attempts to think through the connections between relationality and control.

The figure of twin telepaths occupies a particularly important position in this archive for two reasons that will be the focus of this chapter. The first is the central role of twins in genetics research and their resulting popular association with eugenics and the technological control of life. Conceptions of inheritance underwent a transformation during the post-war period that was closely bound up with the cybernetic perspective on life as information. The trope of telepathic twins mixes a diachronic axis of control (twins as clones) to a synchronic axis (telepathy as mind control). The second reason is the association between twins and myth. 'Scenes from the Life of a Double Monster' ends with the narrator imagining himself and his brother being spotted on a hillside attempting to escape their captivity. '[H]e would have surely experienced a thrill of ancient enchantment to find himself confronted by a gentle mythological monster in a landscape of cypresses and white stones.'[17] So often employed in the construction of mythic national foundations, the cybernetic era has turned to the figure of telepathy for a form of anti-foundational mythology: a mythology not of foundations and national essences but of relations and processes. Narrative has always played an important part in the popularization of telepathy and other 'psionic' abilities, from the case studies presented by the SPR researchers to the use of science fiction by John W. Campbell, the editor of *Astounding Science Fiction* magazine, to, in his words, 'nudge interest in psionic powers as an engineering value'.[18]

The narratives about telepathic twins explored in this chapter, I will argue, function as what David Burrows and Simon O'Sullivan describe as 'mythotechnesis' and define as 'practices which instantiate or perform [...] different human-machine relations'.[19] In his account of the 'invention' of the concept at the end of the nineteenth century, Roger Luckhurst points out that '[t]he conceptualization of telepathy in fact defines its own mode of discursive interconnection: it sparks across gaps, outside recognized channels, to find intimate affinities in apparently distant discourses'.[20] Through its various iterations in cybernetic science fiction, the figure of telepathic twins enables a similar process of cross-discursive sparking. Telepathic twins function as a technology of what Guattari describes as the 'transversal' thinking required to explore the limits of environmental control.[21] After exploring the use of twins as a multivalent metaphor in early cybernetics theory, this chapter will examine three iterations of the use of the telepathic twins trope as part of a mythotechnesis of cybernetics from the fiction of A.E. van Vogt in the 1940s to the work of Philip K. Dick, running from the 1960s to his death in 1982. While Grey Walter and Norbert Wiener alternate between using twins as a metaphor for entropy and negative entropy, the science fiction writers explore the potential of the latter and present twins as open systems that are productive of new configurations of the human and the non-human.

Elmer and Elsie

The first cybernetic twins appeared on a BBC news segment that aired in 1951. Elmer and Elsie were two autonomous robot 'tortoises' described by their constructor, neurobiologist William Grey Walter, as the originating members of a new inorganic species or *Machina speculatrix*. The presentation of the pair of almost identically built 'sibling' robots in the programme contains the tensions and ambiguities present more generally in the use of twins by early cyberneticists to exemplify their theories and processes. Throughout the segment, the alluring strangeness of the animate machines is set off against the mundane familiarity of domestic life. An establishing shot shows us Grey Walter's home, 'a simple villa on the outskirts of Bristol', against the background of a jaunty string and muted trumpet musical score that would be more appropriate in a Laurel and Hardy sketch than for reporting a major breakthrough in robotics. We move quickly to an interior living space where the scientist and his wife smilingly look on as their 'pets' Elmer and Elsie scoot around the carpet in their signature arcing motions, avoiding obstacles and returning to a charging pod when their batteries run out. The voice-over explains: 'they are small and he doesn't dress them up like men. He calls them "tortoises." And so, cunningly, have their insides been designed that they respond to the stimuli of light and touch in a completely life-like manner'.

The emergence of electronic life is presented as a comic interlude in suburban post-war existence. The uncanny is offset and tempered by the familiar and twinship is evoked in service of both. Elmer and Elsie's interactions are described as the familiar competitive behaviour we would expect of siblings. But the fact that the robots are not-quite replicas also evokes the uncanniness of twins. The difficulty in distinguishing between the two tortoises mirrors the main intended effect of the robots as a technological performance: namely, to challenge dominant distinctions between the organic and the mechanical, between human and non-human, between life and non-life. Hayles underlines the importance of electromechanical devices such as Grey Walter's tortoises (alongside Claude Shannon's electronic rat and Ross Ashby's homeostat) in displaying the core ideas of cybernetics at work. 'These artifacts functioned as exchangers that brought man and machine into equivalence; they shaped the kinds of stories that participants would tell about the meaning of this equivalence.'[22] In other words, the uncanny charge produced by Elmer and Elsie's performances, which drew on a long narrative tradition that associates twins with the uncanny, was a keystone in the founding mythotechnesis of cybernetics.

The tensions evident in the BBC report are echoed in Walter's explanation of the mechanics of Elmer and Elsie's interactions with their environments in his popular 1953 book *The Living Brain*. Equipped with an optical receptor that is attracted to moderate lights but repulsed by bright lights the tortoises endlessly scan their environment in search of luminescence. They are also fitted with flash-lamp bulbs on their 'heads' which are turned off automatically whenever the photocell receives an adequate light signal. So, when the light source that they have found becomes too bright as they approach it, their own light switches off. An oscillator then kicks in and they move away to resume their search elsewhere. When both Elmer and Elsie are at work in the same environment, they become attracted and repelled by each other in an endless loop that reproduces their behaviour in front of a mirror. As Walter describes it, '[t]he creature therefore lingers before a mirror, flickering, twittering, and jigging like a clumsy Narcissus'.[23] When two identical robots are attracted to each other's light, 'both extinguish the source of attraction in themselves in the act of seeking it in others. Therefore, when no other attraction is presented, a number of the machines cannot escape from one another; but nor can they ever consummate their "desire."'[24] Walter cites this glitch as evidence of the development of unprogrammed behaviour through their interaction with the external world. In other words, he cites it as evidence of the equivalence in the functioning of organic and non-organic organisms.

Since it is attempting to appeal to the broad popular readership that was the target of Penguin's Pelican Books imprint, *The Living Brain* employs a similar popularizing narrative strategy to the BBC segment. Elmer and Elsie's 'flickering' interaction is framed in terms of that inalienable human emotion, desire, in a way that is in equal measures domesticating (since it is familiar) and jarring (since it is being ascribed to a robot). The reference to the Narcissus myth, meanwhile, lends an atmosphere of monumentality to the description of the behaviour being held up as the first tentative steps taken by a new species. It also uses the rhetorical trick, common in the writings of early cyberneticists, of presenting the emergent behaviour of electronic organisms as non-modern. Rather than the culminating end point of technological development, the robots take us back to the realms of myth, before the enforcement of modern divisions between the animate and inanimate, self and other. As Andrew Pickering puts it, Elmer and Elsie's cybernetic 'performance', looping around each other until their batteries run out and entropy engulfs the endless jittering relay of information, stages 'a nonmodern ontology' by 'engag[ing] directly, performatively and nonrepresentationally with the environments in which they find themselves'.[25] Grey Walter turns to a tale about twins to construct the nonmodern foundations of a cybernetic inorganic organism.

Twins also feature prominently in Norbert Wiener's popularization of cybernetics theory in his 1950 book *The Human Use of Human Beings*, published two years after the more technical *Cybernetics: or Control and Communication in the Animal and the Machine*. The fifth chapter sets out to explain his vision of organic life in terms of informational pattern rather than essence or soul. The growing differences that develop between monozygotic twins are used to exemplify the idea that 'the biological individuality of an organism seems to lie' not in 'the matter of which it is made' but rather 'in a certain continuity of process, and in the memory by the organism of the effects of its past development'.[26] The functioning of the organism in this sense is mirrored by that of the 'computing machine', the behaviour of which is determined by 'the retention of its earlier tapings and memories'.[27] To exemplify this, Wiener draws a comparison between two computers that are constructed to the same specifications and use the same software and human identical twins. '[J]ust as the future development of these two machines will continue parallel except for future changes in taping and experience, so too, there is not inconsistency in a living individual forking or divaricating into two individuals sharing the same past, but growing more and more different'.[28] The proof is that 'this is what happens with identical twins'.[29] Here, the emphasis is placed on the differences that develop between identical twins as a result of epigenetic factors produced through the differing nature of the encounters with their environments. Just as two identical computers will develop differently according to the information inputted into their systems, so the feedback produced through contact with the environment will set twin siblings on differing developmental paths.[30]

The chapter ends by invoking twins as a metaphor for a second time. In this instance, it is the sameness of twins that is emphasized. 'When one cell divides into two', Wiener writes, 'we have a separation in matter which is conditioned by the power of a pattern of living tissue to duplicate itself'.[31] He goes on to claim that this duplication (a process that is most strikingly evidenced by monozygotic twins) proves that 'there is no absolute distinction between the types of transmission which we can use for sending a telegram from country to country and the types of transmission which at least are theoretically possible for transmitting a living organism such as a human being'.[32] Here, Wiener is anticipating the influence that cybernetics was to have on molecular biology during the 1950s. As Lily E. Kay puts it, '[t]hrough the introduction of terms such as *information, feedback, messages, codes, alphabet, words, instructions, texts,* and *programs,* molecular biologists came to view organisms and molecules as information storage and retrieval systems'.[33] The phenomenon of heredity in particular was 'redescribed within [the] system of metaphors, models, and semiotics derived from information theory and cybernetics'.[34] In the words of François Jacob, through the influence of Wiener, heredity was seen to 'function like the memory of a computer' and was conceptualized as 'the transfer of a message repeated from one generation to another'.[35] As the main mechanism for studying heredity, twins became incorporated into this cybernetic 'system' of metaphors and were seen as information machines, crucial points of intersection between individual bodies and underlying genetic codes. In the post-war period, through the strong association between twin studies and Nazi eugenics, twins were seen as a mechanism for technical interventions into the inter-generational 'transfer' of messages. Although it is commonly thought that twin studies fell into abeyance during the 1950s (because of the lingering association with Nazi eugenics), it has become increasingly clear that genetic studies using twins continued unabated in this period, albeit in an unpublicized and somewhat unregulated manner. The 2018 documentary *Three Identical Strangers,* which tells the story of triplets who were separated at birth in the early 1960s and reunited in the 1980s, shines a light on the experiments on twins carried out by psychologist Peter Neubauer throughout the 1950s and 1960s. The focus of Tim Wardle's film is the human cost of these twin experiments: the psychological fallout of their use as scientific instruments (all three brothers were prone to depression and one committed suicide). The unethical treatment of the brothers is foreshadowed by the metaphorical reduction of twins to information that ends the fifth chapter of *The Human Use of Human Beings* and that runs through the cybernetic imaginary of this period.

If the first reference to twins in chapter five of *The Human Use of Human Beings* functions to highlight the importance of embodiment and the agency of embodied interactions with the world, the second effaces embodiment by reducing human life to information. This image of 'transmitting' a human subject across telegraph wires foreshadows the transhumanist thought of Hans Moravec and others who envisage the human as a set of informational processes and resultingly, as Hayles claims, 'rewrite' disembodiment into the 'prevailing concepts of subjectivity' at the very technological and cultural moment when the liberal humanist subject is most vulnerable and open to critique.[36] So, the ambivalence in the treatment of twins in Grey Walter's cybernetic performances (a tension between the uncanny and the familiar) is echoed by a similar tension in Wiener's use of twins as a metaphor for cybernetic organisms. Wiener hesitates between using twins to exemplify human subjectivity as disembodied informational pattern and the processual by-product of embodied interactions with material environments. The contradictions in his use of twins points to the wider tensions in Wiener's work as he tried to reinscribe his cybernetic vision within the dominant liberal humanist concept of subjectivity.

The ambiguity in the use of twins in the metaphorical systems set in motion by Grey Walter and Norbert Wiener is echoed by a further tension in the role of twins in the wider cybernetic imaginary. In cybernetics narratives, the relationship between twins has embodied the dynamics of both entropy and negative entropy. By being made to embody both, twins constituted a key conceptual device for thinking through the social and philosophical consequences of the dominant definitions of information developed by early cyberneticists. Wiener famously defined information as being synonymous with negative entropy. In his conceptualization of information, Wiener drew on the second law of thermodynamics, which states that a closed system tends towards increasing randomness and hence entropic decline, whereas an open system, characterized by ongoing interactions with its environment, becomes more ordered with time. The noise produced by closed systems contrasts with the capacity of open systems to produce information. Wiener's first reference to twins aligns them with the processes of negative entropy. The description of twins as diverging entities, the differences between which increase in line with the feedback they receive from encounters with their environments, positions twinship as an open system. This framing of twins in terms of negative entropy fits within the long tradition of seeing twins as windows onto the determining agential role of genetic information over human behaviour. Twins, in other words, became living metaphors for the vision that emerged in the wake of the Second World War that saw the world, as Hayles puts it, 'as an interplay between informational patterns and material objects'.[37] The overlap in popular culture between the tropes of clones and twins (an association for which *Brave New World* provides the paradigm) is indicative of the way the latter have been taken to embody the view of human life in terms of the replication of informational patterns. Twins continue to provide metaphors for more recent conceptions of the interaction between informational patterns and material objects. In an advert for the consumer credit reporting agency Experian, aired in 2018, the actor Marcus Brigstocke plays a pair of twins who present themselves as 'Dan and Dan's data self'. 'Dan' introduces his twin as a 'physical manifestation of my financial history... made up of things like my transactions, my phone contract, my credit score, stuff like that'. Since 'Dan's data self' is central to determining whether Dan can take out a loan or a credit card, he plays an influential role in Dan's existence. The advert shows the data twin interfering in Dan's life in a number of ways, sometimes obtrusively (attending a romantic dinner with his wife) and sometimes positively (helping them move into a bigger house). The relationship between the twins is used as a metaphor for the interplay between our physical selves and our data selves and as a device that encourages the viewer towards a more stable conception of the links between the two.

This framing of twins in terms of negative entropy stands in tension with a longer tradition that describes the relationship between twins as a closed system that produces the conditions for entropic decline. Grey Walter presents twinship as a closed entropic system in his performance of the uncanniness of electromechanical cybernetic organisms. The endlessly looping 'Narcissistic' behaviour of Elmer and Elsie when they confront each other is compared with the doomed desire for the same often associated with twins. Since Elmer and Elsie are described by Grey Walter as brother and sister, we can assume that he is referring not to Ovid's narrative of Narcissus but to the version developed during the second century AD by Pausanias in his *Description of Greece*. In this less common version of the story Narcissus has a twin sister. When she dies, he assuages his grief by gazing into the water of a spring where he would see 'not his own reflection, but the likeness of his sister'.[38] The Pausanias version ends in the same manner as Ovid's tale, with Narcissus so enraptured by his own image that, in his entrancement, he withers away and dies. Grey Walter presents Elmer and Elsie's mutual entrapment as a cybernetic version of the Pausanias myth. Their dance of unconsummated desire traces a trajectory of entropic decline that ends when their batteries run out. This framing of twins as a closed system doomed to entropic death is echoed through the strong association in gothic literature between twins and degeneracy, a connection that is frequently reproduced or parodied in more recent horror fiction, some of which are the subject of Chapter 2. If Wiener's description of twins as diverging machines positions them in terms of negative entropy and therefore information, the uncanny tradition associates twins with entropy and the noise produced by the breaching of boundaries between self and other, animate and inanimate.

Psience fiction and the limits of Eugenic control

Many of the tensions evident in the metaphorical use of twins by Grey Walter and Norbert Wiener are played out through the trope of telepathic twins as it develops in science fiction narratives from the 1940s to the present. In his account of what he calls 'psience fiction', Damien Broderick traces the evolution of a tendency within science fiction of this period to explore the dramatic potential of extra-sensory perception, a term (often shortened to ESP) that was introduced by botanist and psychologist Joseph Banks Rhine to refer to a range of paranormal abilities including telepathy, clairvoyance, precognition and psychokinesis. Although, as Peter Lowentrout points out, ESP or psionics (to use an alternative term coined by Robert Thouless and Bertold Wiesner) has often been used 'as a displacement of religious concerns', I would argue that the popularity of these fictional tropes coincides with the prominence in public discourse of cybernetics. Although predating Wiener's work (Broderick traces the roots of the subgenre to Olaf Stapledon's novels of the 1930s), psience fiction became an important channel through which cybernetic theory diffused into the popular imagination. The various abilities that are grouped beneath the labels ESP and psi are united in the violence they perform on the boundedness of the liberal human subject. While narratives about telepathy present cognition as distributed among human and non-human subjects, psychokinesis constructs human subjects as developing through intrarelation with the object world. In descriptions of these two abilities (the two most closely associated with twins) cognition and agency are both framed as relational phenomena.

In novels such as A.E. van Vogt's *Slan* (1940) and Edward Elmer Smith's 'Lensman' series, published in the mid-1960s, telepathy in particular functions as a metaphor for the social and psychological impact of emerging electronic communications technologies. One of the main tendencies within the subgenre identified by Lowentrout is 'biological radio'. The future humans in *Slan*, for instance, have developed telepathic powers thanks to their activation of 'formerly little known formations at the top of the brain, which, obviously, must have been the source of all the vague mental telepathy known to earlier human beings'.[39] For the posthuman 'slans' these formations have evolved into organic radio antennae. As Broderick puts it, telepathy is treated 'as if cell phone circuitry had been implanted into the brain'.[40] The association with electronic communications devices is also common in descriptions of telepathy in non-fictional contexts. An intriguing example, and possible direct influence on Stapledon and van Vogt, is Upton Sinclair's *Mental Radio*, an investigation by the author into his wife Mary Craig Sinclair's telepathic abilities, first published in 1930. This connection between ESP and communications technology can be traced back to the mutual influence between spiritualism and communications technologies during the nineteenth century. Heiki Behrend and Martin Zillinger observe that 'spirit mediums and technical media have provided each other with metaphors'.[41] While, as Jeffrey Sconce and others have pointed out, spirit mediums used the latest communications devices and strategies in their performances – from photographs of dead loved ones to telephone conversations with ghosts – attempts to describe the effects of electronic media have often resorted to the language of spiritualism.[42]

In science fiction narratives from the 1940s and 1950s, however, the trope of telepathy took a specifically cybernetic turn. Perhaps the best-known treatment of the telepathic twins trope in science fiction of this period is Robert A. Heinlein's *Time for the Stars*, published in 1956 as part of the publisher Scribner's 'Juvenile' series. The novel combines a fictionalization of physicist Paul Langevin's 'twin paradox' with a speculative narrative about the technological potential of twin telepathy. The twin paradox refers to a thought experiment around the differing rates of aging between two twins if one goes on a round trip to space at the speed of light while the other remains on Earth. In *Time for the Stars*, it is the narrator Tom who travels to space as part of an exploratory voyage to neighbouring star systems directed by the Long Range Foundation (LRF) while his brother Pat stays as home. Since Pat ages much quicker he is long dead by the time Tom finally makes his return. The novel takes place in a version of the future in which, due to their telepathic abilities, 'twins got to be suddenly important'.[43] The significance is due mainly to the discovery that the telepathic link between certain twins retains the strength of its signal even over long distances, making them the most reliable communications technology with which to maintain contact between LRF's Earth headquarters and the crews of the voyaging 'torchships'. Tom first becomes aware of he and his twin Pat's social value when their family is visited by a representative of the LRF's 'Genetics Investigations' scientific agency leading a project intended to 'gather data concerning twins'.[44] In a way that draws on deep cultural associations between twins and eugenics, Tom and Pat become the object of scientific scrutiny and are subjected to a range of biometric tests to ascertain their suitability for the latest mission.

Once they pass the tests with flying colours, Tom and Pat assume their technological roles as communications devices not only between Earth and the latest torchship mission (telepathy is used as a metaphor for communications technology) but also between the current and future generations. The LRF's slogan, 'Dedicated to the Welfare of our Descendants', is indicative of how it views its mission of finding inhabitable planets in other solar systems as a service to the future of humanity. In this way, the novel performs a twist on the role of twins in genetics research. Whereas twin studies use twins as windows onto the influence of the past, due to their capacity to illuminate the agency of inherited factors on human development, the LRF uses them as conduits to the future. While in the opening chapter Tom experiences his close relationship to his brother as a prison from which he longs to escape, once they are separated by the mission, the twin bond is increasingly constructed as an open system. The opening of the twin system is not presented as a consequence of physical distance but of the increasing age gap that opens up between the brothers due to their differing experience of the passage of time. Despite the fact that 'telepathy is instantaneous' and there is no 'signal strength drop off' despite operating over a distance of trillions of miles, Tom starts to experience a 'slippage' between himself and his brother.[45]

> [B]rains are flesh and blood, and thinking takes time... and our time rates were out of gear. I was thinking so slowly (from Pat's viewpoint) that he could not slow down and stay with me; as for him, I knew from time to time that he was trying to reach me but it was just a squeal in the earphones so far as making sense was concerned.[46]

Travelling at a 'gnat's whisker' under the speed of light for just a moment would be the equivalent of ten years 'Earthside'. Following a period of high-speed travel, both twins can still 'telecommunicate satisfactorily' but, from then on, 'it was like two strangers using a telephone'.[47] As the connection with his brother grows weaker, to Tom's surprise he picks up a line of telepathic communication first with his niece Molly and then with his great grandniece Vicky.

The openness of Tom's telepathic connection inaugurated by his twin bond, aligns *Time for the Stars* firmly with Norbert Wiener's construction of twinship as a metaphor for information as negative entropy. Tom's psychic life mutates in unpredictable and unprogrammed ways as he incorporates different 'voices' into his subjective assemblage. These mutations become necessary to maintain his status as a mediating channel between the headquarters on Earth and the travelling torchships. In other words, the twin bond is presented as a communications technology that adapts in response to changes in the environment through a series of feedback loops. So, while from the beginning the twins are equated with communications technologies (Tom and Pat's brains are repeatedly compared to radios), as the narrative unfolds, they are presented as a cybernetic device. This construction of the twins as a model of a cybernetic organism displays a similar tension between the familiar and the uncanny as was evident in the Elmer and Elsie performance and its reception in the media. On the one hand, the fact that Tom himself acts as narrator naturalizes his telepathic abilities. In the opening chapter he paints a realistically complex portrait of his and Pat's relationship. Having a twin brother, he explains, 'makes you close; it makes you almost indispensable to each other – but it does not necessarily make you love him'.[48] The familiarity of this ambiguous dynamic is offset against the uncanniness of Tom's experience. The final chapters recount Tom's return to a hometown that differs considerably from that which he left. The strangeness is emphasized by the final twist when it is revealed that Tom marries his great grandniece Vicky. When Vicky declares her intention to marry her great uncle moments after they have met in person, she explains: 'You seem to forget that I have been reading your mind since I was a baby – and a lot more thoroughly than you think I have!'[49] Here, Heinlein makes an association between twins and incest that is common in gothic narratives of uncanny twins. But, in contrast to the twin siblings in Edgar Allan Poe's 'The Fall of the House of Usher' (discussed in Chapter 2), the twins in Heinlein are not used as a metaphor for degeneration. Rather, the twin bond is presented in keeping with the views of Myers and his fellow SPR researchers for whom, in Luckhurst's words, 'telepathy and other forms of hyperacuity were markers of evolutionary advance, rather than symptoms of reversion'.[50] If, as I am arguing throughout this chapter, the trope of twin telepathy is a key component of the mythotechnesis of cybernetics, then in *Time for the Stars* it is used to align the cybernetic vision with a eugenicist agenda. Twins are presented as a communications technology with which to control the passage of information across generations. However, alongside this reduction of twins to systems of communication and control there are glimpses of more critical uses of the twin metaphor as a model for the processual ontology described by Pickering.

The focus on the connection between communication and control is central to another key interpretation of the telepathic twins trope. A.E. van Vogt's novel *The Weapon Makers* was first serialized in *Astounding Science Fiction* magazine from February to April 1943, reworked and published in book form in 1947 before reappearing in 1952 in thoroughly revised form. The 1943 version centres on the character of Dan Neelan, an atomic engineer on a meteorite mine who returns to Earth when his telepathic connection with his twin brother Gil abruptly breaks. 'Acute awareness it had been of the lack of that neural pressure which had constituted, even at that distance, the bond between his brother and himself'.[51] Fearing that Gil has died, but being unable to find him, Dan embarks on an action-filled search for his lost twin across a far-future metropolis ruled by the technocratic Isher Empire. He soon discovers that his brother, a talented engineer, was part of a team that had invented an atomic rocket set to revolutionize space travel, control over which is the object of a ferocious struggle between the Isher rulers and their main opposition, an elitist libertarian guild called the Weapons Shop, characterized by 'an independent, outlawed, indestructible, altruistic opposition to tyranny'.[52] Fredric Jameson argues that, together with Heinlein and others, van Vogt's work of the 1940s contains many of the characteristics associated with the Golden Age (a focus on space adventure often with a techno-utopian ethos) while paving the way for the likes of Philip K. Dick with his interest in psychoanalytic frameworks.[53] van Vogt's innovations lie not in a virtuosity of style or strikingly original premises but in the reworking and reshuffling of existing pulp conventions. His treatment of telepathy through the characterization of the Neelan twins repositions the trope within the proto-cybernetic imaginary of the 1940s. Whereas *Time for the Stars* incorporates its telepathic twins into Heinlein's techno-utopian eugenicist agenda, van Vogt employs the trope to trace the limits of eugenic control.

Dan and Gil's telepathic bond is presented as at least partly the result of techno-scientific intervention. Widowed not long after her sons were born and unable to afford their education, the twins' mother 'puts them under contract' with the state eugenics institute.[54] Already 'sensitive to each other,' the scientists at the institute 'magnified the sensitivity until it was a warm interflow of life force, a world of dual sensation'.[55] The bond between the two brothers is modified and tweaked like the latest electronic communications device. By the time their contract ends and the twins go their separate ways, 'the interrelation had grown so sharp that at short distances – a few thousand miles – thoughts passed between them with all the clarity of the electronic flux in a local telestat'.[56] The experimental training that the twins undergo while under contract with the Institute contains, in a condensed time scale, the role of twin studies in eugenics research. What eugenics projects aim to do over the course of generations van Vogt's Institute achieves in a matter of months. In the future world of *The Weapon Makers*, the twins are not the only humans to be the object of technological control. The dynastic line that governs the Isher Empire, whose dominance has guaranteed peace for hundreds of years, has perfected a system of mind training that ensures their intellectual and physical superiority over their subjects. The vision of technological control developed in van Vogt's fictional universe is completed by a complex biometric surveillance system. The posthumans in *Slan*, for instance, are regularly subjected to a barrage of tests, including 'fingerprinting, X-ray, blood test, chemical reaction of the skin, microscope measurement of hair, and so on'.[57] In *The Weapon Makers*, when Isher security agents attempt to pre-empt the actions of a spy working for the Weapons Shop guild, they build electronic models of his brain, what they describe as 'cerebro-geometric figures', to predict his thought patterns. The Isher scientists maintain an elaborate system of information about their potential enemies: 'graphs of brain and emotional integers, curious mathematical constructions whose roots delved into the obscure impulses of the human mind and body'.[58]

The biotechnical engineering that is routine on Earth in *The Weapon Makers* is thrown into relief by that of the alien species with which Dan comes into contact in the final chapters of the novel. When he hijacks the atomic rocket ship that his brother designed and launches into space, Dan is taken prisoner by a mysterious and hyper-intelligent species that communicates with him and with each other telepathically. Although their physical form is described as spider-like, the alien intelligence is characterized as an amorphous, ever-shifting organism that incorporates external entities in an ongoing autopoietic process. They tell Dan that 'on entering our magnetic field your ship became a part of it. It has already been dissolved into a complex screen and will form part of our organized reserve of developed energies'.[59] Their chance encounter with the lost spaceship has prompted them to incorporate humanity's systems of cognition into their own. And the one aspect of human communication and knowledge that jars with theirs, and which they are therefore more interested in understanding, is the capacity for empathy that they find in its purest state in the telepathic bond between twins. So Dan finds himself once more the object of scientific study. To carry out their 'investigation of man's sensory equipment' they locate Gil's body where it lies on a nearby planet in the Alpha Centauri star system and bring it back to life and Dan experiences this resurrection from a distance. 'They were together now, Gil and he; he could feel his brother's slow heart-beat, could see all the mind pictures that pushed hesitantly into a brain that had scattered far'.[60]

The proto-cybernetic nature of the alien entity's worldview is clear. Their fascination with the Neelan twins is indicative of their interest, not in discrete autonomous human entities, but in the intrarelational processes that bring them into being and define their parameters. Emotion, as the aliens describe it, 'is an energy' that 'acts instantaneously over a distance' and occurs between individuals.[61] Here, the aliens develop an account of telepathic communication that differs significantly from that of the Eugenics Institute which compared it to the functioning of a 'telestat'. The emotional communication that takes place in twin telepathy, as the aliens see it, is far noisier than this metaphor suggests. It is the noisiness of the connection that eludes the control of both the alien intelligence and the Isher security agents. The telepathic link that exists between the Neelan brothers is emphasized as an embodied form of communication. When he arrives back on Earth, Dan regains a degree of hope that his brother is still alive when he starts to feel a 'sensory connection' establishing itself at the limits of his conscious mind: 'Tenuous, imperceptible during waking hours, it had manifested itself to that supersensitive structure that is the human body during sleep hours'.[62] When the connection is severed it is experienced as physical pain: 'The break was so sharp, it hurt like fire.'[63]

This account of telepathy as an embodied form of communication, a quality captured by the etymological roots of the term, is repeatedly emphasized in popular accounts of twin telepathy. Playfair, for instance, points out that in the vast majority of narratives, telepathic connections between twins are triggered by moments of intense physical pain or emotional distress.[64] But the emphasis on embodiment in *The Weapon Makers* is also in keeping with Wiener's partly disavowed critical posthumanist description of twins as cybernetic organisms that develop separately through distinct embodied interactions with their environments. For van Vogt, the telepathic twin bond entails an expansion of Dan and Gil's embodied interface with the world that does not preclude development along different paths. Dan and Gil's twin bond is presented as very much an open system and one that does not fit either the aliens' concept of intelligence or the biometric control systems employed by the Isher regime. This openness is emphasized by van Vogt's deployment of doppelgängers in the text, which occurs alongside the trope of twinship. When Dan's spaceship is first incorporated into that of the alien, Dan steps out of his vehicle and finds himself on a version of Earth that is familiar from his memories but also irrevocably strange. His suspicions that it is just an imitation of Earth ('such a perfect imitation that – It couldn't be anything but an inhumanly marvellous set-up') are confirmed when he finds another version of himself in his spaceship replaying an incident from his earlier life. Confronted by this jarring spectacle, Dan calmly reminds himself that it is part of the alien experiment. 'The feel of the clothes was unmistakable; and the flesh of the face was warm with life as he touched it with his fingers.'[65] But the experience does not arouse the usual horror we would associate with the doppelgänger. Dan is alert to the possibilities of machinic replication and the porousness of the boundaries of his selfhood. Within the logic of the narrative, Dan's twinship has paved the way for the schizophrenic unbinding of the self marked by the intrusion of his double. In fact, the reader comes to suspect that the alien itself is also a double of the engineer projected outside of himself in a moment of grief for his brother. Its amoeba-like structure acts as a mirror for the autopoietic assemblage of which he is a part. So while at the beginning of *The Weapon Makers* the twin trope is used as a metonym for biopolitical systems of control, it ends by opening out onto a hallucinatory vision of mutation and becoming. While both writers discussed in this section are interested in twin telepathy as a conceptual device for analysing the connection between communication and control, the focus of van Vogt is on the limitations of control.

Superorganisms and asignifying rupture

The positioning of the twin bond within an open cybernetic system is taken to an extreme by a novel published two years prior to *Time for the Stars*, in 1954: Theodore Sturgeon's *More Than Human*. The narrative, which found its first readers as a short story in *Galaxy* magazine before being expanded into novel form, tells the story of a group of outcast children and teenagers who discover that they have psionic powers that include telepathy, telekinesis and computational intelligence. When they come together, these forces enable them to form a superorganism that incorporates and activates the abilities of the individual components. These components include telepaths Lone and Gerry; the telekineticist Janie; an autistic super-intelligent infant, referred to as Baby; and monozygotic twins Bonnie and Beanie who, though almost completely mute, are able to teleport themselves across space, the confusing effect of which is redoubled by their startling physical likeness. Gerry, who becomes a self-appointed leader and chief theorist, describes the group as the first of an emerging species, *homo gestalt*, set to replace *homo sapiens* as the 'next step in evolution'. The concept of the telepathic superorganism was not particularly original when Sturgeon was writing. John Wyndham used the idea in two of his novels from the 1950s – *The Chrysalids* (1955) and *The Midwich Cuckoos* (1957) – and the trope has since been developed in Octavia Butler's 'Patternist' series (1976–84) and more recently in the television show *Sense 8* (2015–18). But *More Than Human* constitutes a particularly interesting moment in the development of the science fiction superorganism for the explicit dialogue it sets up between the history and theory of cybernetics and experimental psychotherapeutic practices such as gestalt therapy that would become central to the psychedelic vision later developed by Philip K. Dick (the subject of the final part of this chapter). It is

Sturgeon's anticipation of this vision that is picked up on and enhanced in the graphic novelization of *More Than Human* published by *Heavy Metal* comics in 1978 with an adapted script by Doug Moench and art from Alex Niño.

The history of cybernetics is woven into the narrative of *More Than Human* in a way that highlights its relevance as a conceptual framework for its 'homo gestalt' protagonists. The dramatic tension in the novel derives largely from Gerry's attempts to hide the existence of their group from a military engineer named Hip Burrows. Burrows is first alerted to the group's activities when he finds a mysterious machine half-buried in a firing range used for training anti-aircraft gunners that seems to inhibit magneticism. Burrows immediately sees the military potential of the discovery ('the perfect defence mechanism for the electromagnetic age'), but his investigation is stymied before he can prove its significance.[66] The machine, it emerges, is an anti-gravitational device produced, almost unwittingly, through the collaborative efforts of Baby and the 'idiot' but able-bodied telepath Lone. Gerry hides the machine, but not before it has enraptured Burrows with its revolutionary potential: 'Antigravity, which would change the face of the earth in ways which would make the effects of steam, electricity, even nuclear power, mere sproutings of technology in the orchard this device would grow'.[67] The technologies that the antigravitational device disrupts are the antiaircraft systems conceptualized by Norbert Wiener during the Second World War and through which he developed his theories of cybernetics. As Peter Gallison explains, 'Faced with the problem of hitting fast manoeuvrable bombers with ground-based artillery, Wiener brought to bear his own established interest in feedback mechanisms, communication technology, and nonlinear processes.'[68] This reference to the origins of cybernetics is reinforced by a sequence describing a fairground ride in which Hip and Janie operate 'a couple of war surplus servo-mechanisms rigged to simulate radar detectors'.[69] When Janie misses, Hip reminds her: 'You don't have to hit 'em these days, you know. Just get near. Your fuses take over from there'.[70] So, on the one hand, the 'homo gestalt' is described as being cybernetic in nature: a complex assemblage of communications and feedback devices that allow its participants to 'blesh', as they put it, with the twins functioning as the key points of mediation, 'messengers between Gerry and Baby'.[71] On the other hand, the superorganism is presented as an entity that disrupts the military ideology of cybernetics as a science of control.

The other main conceptual framework employed in the novel, psychotherapy, undergoes a similar fate as that of cybernetics. As a mechanism of control through the enforcement of normative behaviour, it fails, having proven insufficient in the face of the emergent superorganism. The second section of the novel is narrated through a dialogue between Gerry and a psychotherapist called Stern. The session starts with Gerry confessing to murder and during the dialogue he is encouraged to work through his memory of events leading up to the crime, including his involvement with Jeanie, Baby and the twins Beanie and Bonnie. The main tension of the section is provided by the conflicting aims of the analyst and the analysand. While the psychotherapist encourages his client to identify his own individual motivations, in isolation from the group, Gerry uses the session to understand the group dynamic with greater clarity. Rather than play out the drama of the private self, the session serves as a springboard for the rise of the networked self of the '*gestalt* organism'. The section ends on an ambiguous note with Gerry on the one hand predicting a future ruled by collective organisms (and performing the experience of 'bleshing' during telepathy) and, on the other hand, violently asserting his own individualism as leader ('I'm the central ganglion of a complex organism').[72] This paves the way for the more complete realization of the networked organism in the final chapter when Gerry gives up his dreams of private power and merges with the network. The psychoanalytic process, in other words, gets derailed into a dynamic that more closely resembles what Guattari terms schizoanalysis. Stern's discourse acts as a foil for the emergence of what Guattari describes as 'the modalities of "group-being" [*l'être-en-groupe*]'.[73]

The role of telepathic twins is central to this emergence. In a description of the 'mutations' necessary to create the conditions of what he calls 'social ecosophy', Guattari argues that 'at the heart of all ecological praxes there is an a-signifying rupture, in which the catalysts of existential change are close at hand, but lack expressive support from the assemblage of enunciation'.[74] The clash between Gerry and Stern in the psychoanalysis session frames the emergence of the gestalt organism as just such an a-signifying rupture. The phenomena that Gerry is describing defy the categories with which Stern attempts to explain and therefore control them – a moment of epistemological failure that is mirrored by Hip's thwarted investigation into the antigravitation machine. The sinking of these conceptual frameworks clears the path for the half-articulated theories about gestalt being voiced by Gerry. The act of fusion itself requires the creation of new categories and neologisms (such as 'bleshing'). But they are also indicated by the process of telepathy itself, which is repeatedly described as operating at the limits of language. The first discovery of telepathic abilities takes place in the opening pages when Lone connects with a previously unknown girl. The communication is a stutter, flickering in and out of comprehension: 'Without words, though. [...] Radiations of fear, tense fields of awareness, discontent. Murmuring, sending, speaking, sharing, from hundreds, from thousands of voices'.[75] The primary role of the twins in the novel is to embody this point of a-signifying rupture. From the moment they are first introduced, Beanie and Bonnie are described as inhabiting a realm of perpetual pre-linguistic infancy and they seem to incorporate others into this wordless world. The first day the telekineticist Janie spends with the twins 'she spoke hardly a word [since the] twins had not yet learned to talk'.[76] However, this barrier 'was incidental to another kind of communion'.[77] Through this 'communion' Janie experiences a 'sudden opening' that provides the first spark of the gestalt organism.

From these wordless beginnings, the narrative traces a tension between attempts to articulate this communion and the resistance of the collective organism to explanation through existing categories. This tension is presented as a correlate for a further ambiguity that runs through *More Than Human*: whether incorporation into the gestalt organism is a form of liberation or a further form of repression. It could be argued that the double existence of the protagonists – their existence as individuals prior to incorporation and their collective being afterwards – stages what Maurizio Lazzarato describes as the double logic of power under informational capitalism, which echoes the alternation between individuality and 'dividuality' outlined in the Introduction. Lazzarato's double logic functions through a 'two-fold cynicism': 'the "humanist" cynicism of assigning us individuality and pre-established roles (worker, consumer, unemployed, man/woman, artist, etc.) in which individuals are necessarily alienated; and the "dehumanizing" cynicism of including us in an assemblage that no longer distinguishes between human and non-human, subject and object, or words and things'.[78] The 'production of subjectivity' takes place at the intersection of these two dynamics, between 'social subjectivation' and 'machinic enslavement'.[79] From this perspective, it could be argued that the role of Beanie and Bonnie in the gestalt organism is a form of machinic enslavement that is constitutive of the production of capitalist subjectivity. The narrative requirement to continue referring to each member of the group as individuals alongside the descriptions of group being could be cited as further evidence of this double logic.

Lazzarato is employing the term 'enslavement' as it was developed by Gilles Deleuze and Félix Guattari in explicit dialogue with cybernetics: 'It means the "management" or "government" of the components of a system. A technological system enslaves ("governs" or "manages") variables (temperature, pressure, force, speed, output, etc.), ensuring the cohesion and equilibrium of the functioning of the whole'.[80] In this system of cybernetic enslavement, the relations that exist between humans and non-humans are not representational. Rather, 'Human agents, like non-human agents, function as points of "connection, junction, and disjunction" of flows and as the networks making up the corporate, collective assemblage, the communications system, and so on.'[81] From this perspective, the psionic powers that join the members of the group and continually defy the limits of language can be seen as metaphors for these asignifying processes of junction and disjunction. Gerry, the leader of the group, is the cybernetic steersman (from the Greek 'kubernetes') who directs the flows of the system. While social subjection operates through 'binary segmentarities', enslavement functions with a 'flexible segmentarity, which traverses subjections and their binaries'.[82] The race and gender dynamics in *More Than Human* also seem to follow this logic. The origin stories of each member of the group described in the first section of the book emphasize their status as social outcasts. Before the discovery of the gestalt organism, all the children have been excluded from full social participation for reasons of gender, race, poverty or disability. The social exclusion of the twins Beanie and Bonnie functions at the intersection of all these categories. They are poor (their father is a janitor who beats them), black (and at several moments in the narrative the victims of racial abuse), and disabled (their silence, which is at first presented as being due to their age, seems to develop into adult muteness). The fact that they are twins and not fully individuated as separate characters in the novel underscores their exclusion not only from dominant social categories but also from the category of the human itself. Their social exclusion in everyday life contrasts with their very active participation in the gestalt organism. However, even though the racial and gendered biases are not a factor with the group, Beanie and Bonnie's actions are just as restricted by the functioning of the group as a whole. In fact, arguably the desubjectification that they suffer in the social sphere has prepared them for the machinic enslavement that they suffer within the cybernetic organism.

However, the real critical potential of the novel is its staging of the parallel functioning of the logics of social subjection and cybernetic enslavement (as well as the signifying and asignifying regimes that support these logics) and the exposure of gaps that open up between these regimes. The processes of formation and dissolution of the organism produce movements of rupture with signifying systems that are not fully captured by and are excessive of the cybernetic organism. The twins Beanie and Bonnie are presented as trickster figures that operate at the limits of both systems. While their ability to teleport themselves defies the laws of Euclidean geometry, they exploit their twinship to manipulate the borders between the visible and the invisible. When Gerry first meets the twins, he thinks they are one person and is confounded by her ability to disappear and reappear in different far-off places. When he asks Janie how they do it, she replies '*It's easy. She's really twins*'.[83] This ability to stage ruptures in signifying systems is mirrored by the textual structure of the graphic novel version of *More Than Human*. The tension between signifying systems is, in many ways, constitutive of comics as a medium. Charles Hatfield argues that tensions between distinct '*codes* of signification' – alongside tensions between '*single image* and image-in-series'; 'narrative *sequence* and page *surface*'; and 'reading-as-*experience* and text as material *object*' – are 'fundamental to the art form'.[84] While all comics negotiate this tension with different strategies and to different ends, in Moench and Niño's adaptation the relationship between word and image mirrors and draws attention to the interplay between social subjection and machinic enslavement.

The most striking formal quality for the reader of the graphic novel version of *More Than Human* is the uneasy relationship between word and image. The distinctions between the two are carefully policed throughout. Words are scrupulously separated from images. The conventions that blur the distinctions between the two – such as speech bubbles and onomatopoeia – are notable in their absence. The effect for the reader is that the transition between the two is also experienced as a break or rupture. This textual awkwardness is, of course, partly the result of its publication history. The novel that provides the source material hangs like a spectre over the book as a whole, reinforcing the sense that authority lies with the written word and inhibiting the development of a distinct autonomous project. However, it also has a critical effect of emphasizing the gaps and disjunctures within the gestalt organism itself. The lack of integration between the two codes opens the possibility of modes of connection between the images themselves that are left unrestrained by textual control. In a way that mirrors the teleportation of Beanie and Bonnie, the reader's eyes are left to make connections across and between pages against the flow of the narrative. In other words, the textual system of the book does not cohere in a way that draws attention to the dissolution of the cybernetic organism described within the narrative. This lack of control shared by both is emphasized within the story by the divergence between the organism and the military roots of cybernetics theory. So, whereas the twins in *The Weapon Makers* stage a cybernetic mode of being that gestures beyond control by opening up to sequences of becoming, *More Than Human*'s cybernetic twins gesture beyond control by emphasizing points of asignifying rupture within informational capitalism.

Philip K. Dick's transitional twins

Twins are central to Philip K. Dick's exploration of cybernetic ontologies and feature prominently, in various permutations, throughout his work. As a recurring narrative figure, twins play a key role in establishing connections between individual psychology and wider social structures, humans and technology, as well as his fiction and the heavily mythologized versions of his own life with which his creative writing is interwoven. Philip K. Dick was born six weeks prematurely with a twin called Jane who died of malnutrition when she was less than a month old and the loss played a prominent role in his personal mythology. The death of Jane was recounted numerous times to journalists and referenced directly in the voluminous writings he produced towards the end of his life in 1982. From a young age, his mother Dorothy would explain to Dick that his sister died despite her best efforts to keep her alive. In the process, she would unwittingly saddle her son with guilt at having survived. He later confided to his third wife Anne that 'I heard about Jane a lot and it wasn't good for me. I felt guilty – somehow I got all the milk'.[85] Biographer Lawrence Sutin claims that: 'The trauma of Jane's death remained the central event of Phil's psychic life. The torment [...] manifesting itself in difficult relations with women and a fascination with resolving dualist (twin-poled) dilemmas – SF/mainstream, real/fake, human/android'.[86]

In placing such narrative weight on the death of Jane, Sutin is falling into a trap of echoing the psychodynamic explanations of Dick's behaviour and preoccupations often favoured by the author himself. Dick had a life-long interest in psychoanalytic theory and would often use his readings to diagnose himself (from 'schizoid affective' in his childhood to 'borderline psychotic' later in life). As Roger Luckhurst explains, the various psychoanalytic explanatory models that were made available to him were not only ways of reframing his understanding of himself but were also actively formative of his behaviour and psychic life. Dick was particularly susceptible to this process of what Luckhurst describes, using Ian Hacking's term, as 'dynamic nominalism'.[87] Dick's founding myth of his lost twin was repeatedly reworked and repurposed in the service of these shifting frameworks. Picking up on these dynamics in Dick's fiction, critics such as Sutin have often diagnosed him as 'a lifelong melancholic, burdened with guilt for surviving his dead twin sister, and left with impossible mourning or forever incomplete individuation, compelled to write fictions of twinning or multiple fragmentation'.[88] Luckhurst claims that Dick's career lends itself to the concept of the 'traumatic subject' favoured by critics in the 1990s: 'an irresolvable loss driving obsessive compulsions to repeat'.[89] Furthermore, the sense of incompleteness left by the death of the twin is projected onto most of the characters who populate his fiction. Almost without fail, their interior lives are characterized by lack: whether it be a lack of humanity that equates them with androids, or living under a diagnostic label that positions them as somehow less than normal. Twins become a point of intersection between individual psychology and the wider social fabric.

Erik Davis adds to this dynamic a consideration of Dick's enduring faith in drugs as a stimulator and regulator of psychic life. By the time he was writing his most influential science fiction during the late 1960s, 'psychoactive chemicals had become so integral to Dick's life and psyche that he had become a kind of pharmaceutical cyborg' who was '*hacking* himself with drugs'.[90] Therefore, the frameworks drawn from depth psychology stood in continual tension with physiological behaviourism, which 'overwrites the hermeneutical operations of psychoanalysis with technical models and chemical solutions'.[91] The recurring trope of psionic abilities in Dick's fiction (telepaths and telekineticists are commonplace in his novels from the beginning of the 1960s onwards) is used to navigate a path between these opposing frameworks, between 'the Scylla of depth psychology and the Charybdis of medical psychiatry'.[92] Just as the concept of telepathy was developed to reconcile materialist with spiritual accounts of life towards the end of the nineteenth century, psionics in Dick's fiction mediates between humanist and posthumanist accounts of subjectivity. It is used as a multivalent metaphor that veers between being used to stand for the irreducibly human nature of empathetic connection (in a way that echoes *The Weapon Makers*) and the constitutive interconnections between humans and technological systems. So, if the twins trope mediates between individual psychology and social structures, telepathy mediates between humanist and critical posthumanist conceptions of human subjectivity. And in Dick's fictional world, all twins are telepaths.

In *How We Became Posthuman*, Hayles places twins at the centre of her account of Dick's intervention into cybernetic conceptions of subjectivity. Her discussion of Dick's work places him in dialogue with 'second order' cybernetics embodied by Humberto Maturana and Francisco Varela's work on autopoietic systems. The characteristic that distinguishes second-order cybernetics from the first-order theories developed by Wiener and John Von Neumann is an interest not only in autonomous systems that moderate their behaviour in response to feedback from the environment, but also in the formative role of the observer in the functioning of these systems. Dick, a 'system builder' like Maturana and Varela, mirrors these concerns through his fascination with the concept that the 'observer creates a system by drawing a distinction between inside and outside'.[93] Dick exploited the narrative potential of this dynamic in narratives that would blur the distinction between inside and outside by revealing how the observer of a system (and hence the entity who maintains the borders of that system) was him or herself incorporated into a system being observed by another entity. Key novels published in his most prolific phase of science fiction production – such as *The Three Stigmata of Palmer Eldritch* (1964) and *Ubik* (1969) – pursue this logic to a point of infinite regress. Others stage a dynamic of radical instability in which a male protagonist, often in the grip of hallucinogenic drugs or schizophrenic breakdown, alternates between being the observer and the observed, inside and outside, human and non-human.

The interpenetration of psychology, technological systems and social structure is central to the exploration of these cybernetic dynamics. As figures that both deploy and undermine binary oppositions between self and other, inside and outside, twins and telepathy (two tropes that are inextricable in Dick's fictional world) are key components in this system. Rather than produce one work centrally concerned with twins, they feature as a key motif that is woven throughout his writing. In an article from 2000, Fredric Jameson argues that the only way to study Dick's fiction from a certain period or 'cycle' is as 'variants of one single work'.[94] Not only is this required by the specific qualities of his fiction (the obsessive repetition of motifs – the product, in part, of the speed with which they were written) but it also encourages a necessary shift in perspective from the 'meaning' of individual works or motifs to the 'representational dilemmas' that are confronted and worked out in the broader fictional system.[95] In an earlier article from 1975, Jameson identifies the inspiration for this strategy as Claude Lévi-Strauss's analysis of myth in terms of 'a narrative construction of symbolic mediations or synthesis whose purpose is the resolution, in story form, of a contradiction which the culture in question is unable to solve in reality'.[96] This systemic framework is particularly appropriate for approaching the role of twins in Dick's fiction. Dick exploited the cultural associations between twins and foundational narratives to use twinship, and particularly twin telepathy, as a central figure in the elaborate mythotechnesis he developed through his fiction. The figure of twins is employed to reshape common conceptions of connections between humans and technology through the construction of twins as transitional objects.

Hayles's analysis of the work of Philip K. Dick in *How We Became Posthuman* focuses on how the confusions between inside and outside the self that recur throughout his fiction are encoded with a 'broader social interrogation into the inside/outside confusions of the market capitalism that incorporates living beings by turning humans into objects at the same time that it engineers objects so that they behave like humans'.[97] The twin relationship is a central figure through which this entanglement is narrated and scrutinized. As Hayles points out, Dick claimed to imagine that he carried his twin sister within himself throughout his life. Jane was thought of as a 'shadowy Other within his body, a figuration that reflected the fact that Jane no longer had an autonomous existence apart from his imagination of her'.[98] In this phantasmatic configuration, Jane 'was fated to occupy the subordinate "inside" position while he, the surviving twin, had an "outside" subject position that made him a recognized person in the world'.[99] With this as the origin story of his twinship, relationships between twins in his fiction are often used to reimagine or renegotiate this configuration of inside and outside. Hayles argues that the female characters with which Dick's male protagonists form relationships can be placed on a spectrum, on one extreme of which is the archetype of the 'dark-haired-girl' while the other is occupied by the 'schizoid woman'. If the dark-haired-girl plays the role of 'anchor in reality' for the troubled male protagonists and is characterized as empathetic and available for human connection, the schizoid woman leads the male protagonist further down the path of the fragmentation of his sense of self and is characterized as cold and disconnected. While in some stories and novels, the different types and the narrative functions they bring with them are performed by separate characters, in others one character will display both tendencies and veer disconcertingly between them.

Hayles claims that Jane was 'the first dark-haired-girl', citing as evidence the fact that Dick often imagined that his sister would have grown into a nice dark-haired-girl.[100] I would argue that the twin bond occupies a less fixed position on this spectrum. On the one hand, the dark-haired-girl is frequently placed in an 'internal' position relative to schizophrenic male characters by occupying a narrative terrain that they regard as more solidly real and therefore confirm a framework of reality that they can readily control. The character of Nicole in his 1964 novel *The Simulacra* provides the paradigm for this type. On the other hand, the schizoid woman can often assume a position of externality, from which vantage point she can control or manipulate the male protagonist's sense of reality and selfhood. The character of Alys Buckman performs this role in relation to her twin brother Felix Buckman and the other male character, TV star Jason Taverner, in the 1974 novel *Flow My Tears, the Policeman Said*. When Alys takes an experimental drug called KR-3 she incorporates those around her into the hallucination, radically warping their sense of reality. Felix and Taverner find themselves playing roles within her drug-addled psyche. Alys's role as a destabilizer is emphasized by the fact that she and her brother have an incestuous relationship, a detail that Felix, who occupies a prominent public position, is keen to keep secret. The drama in the narrative is provided by the male characters' attempts to reassert control and take back the position of externality that Alys has temporarily seized.

The role of twins in framing the struggle between internality and externality is clearest in the 1965 novel *Dr Bloodmoney*, which tells the story of the formation of a community in rural California seven years after the obliteration of North American society by a nuclear war. One of the inhabitants is Edie Keller, a young girl who carries her twin brother Bill, a *foetus-in-fetu*, within her body, with whom she communicates telepathically. In the elaborate formalistic analysis of the novel carried out by Jameson, he argues that the mythic 'character system' developed in the novel is structured around a number of intertwining oppositions, the synthesis of which offers a 'glimpse into a reestablished collectivity'.[101] Within the logic of the novel, the relationship between Edie and Bill provides the blueprint for this nascent social being. In the denouement of *Dr Bloodmoney*, Bill the 'homunculus' leaves Edie's body to occupy that of another character. Hayles claims that the relationship between Edie and Bill needs to be read in relation to Dick's imagined relationship with Jane. At the beginning of the novel 'Edie and Bill are Dick and Jane turned inside out'. The conclusion, meanwhile, 'reverses the tragic end of Dick and Jane's twinship': 'Instead of two children becoming one when the girl twin dies, here one child becomes two when the boy twin succeeds in leaving his sister's body and living on his own'.[102] This 'turning right-side out of a tragically enfolded twin boy and girl' cuts through the 'Gordian knot' formed by the entanglements between self and other, human and technology, that are central to informational capitalism and are figured by the schizoid characters. While I agree with Hayles's assessment of the key role played by the twin bond in Dick's negotiation of these entanglements, I would disagree with this conclusion. Seen in the context of the development of the twin trope through his work as a whole, the picture is not one of neat resolution. Rather, twins are used to establish a temporary equilibrium within the zone of entanglement.

A useful framework here is Bernard Stiegler's appropriation of psychoanalytic theorist Donald Winnicott's concept of transitional phenomena. In his 1951 paper 'Transitional Objects and Transitional Phenomena,' Winnicott discusses the role of objects in the transition of the infant from a state of illusory omnipotence, in which there are no boundaries separating self and world and all its demands are met, to a stage in which the infant experiences itself as separate from an external reality. He argues that the space opened up during this phase, which is no longer part of the self and yet not fully part of an external world, is often embodied by an object, which he therefore labels as 'transitional'. In the introduction to his 1971 book *Playing and Reality*, Winnicott goes on to argue that when this object is 'decathected' (detached from the ego) and therefore loses its meaning, it is because 'transitional phenomena have become diffused [...] spread out over the whole intermediate territory between "inner psychic reality" and "the external world as perceived by two persons in common", that is to say, over the whole cultural field'.[103] Crucial for this analysis is that the 'special feature' of this 'potential space' is that 'it *depends for its existence on living experiences*, not on inherited tendencies'.[104] Technology, for Stiegler, occupies precisely this space of transitional phenomena that are both internal and external to the self. As a supplement that is constitutive of the human subject, technology occupies a position of liminality. As Tania Espinoza puts it, '[t]he domain of transitional phenomena designates in Stiegler's work precisely that region in between the dead, inert mechanical being and the living, organic biological being, where the technical being belongs'.[105] It is clear from this description that twins, in Dick's fictional world, perform the function of transitional objects for each other. Echoing Wiener, Dick breaks with the dominant cultural association surrounding twins to frame the twin bond as the embodiment of a transitional space that depends on 'living experience' rather than 'inherited tendencies'. Twins are rarely associated with genetics in Dick's fiction. Rather, they are evoked to dramatize the intermediate zone between human subjects and between humans and technological systems. Their central function is to reveal this as a space of potentiality.

Twins inhabit the zone between 'dead' technicity and living being in Dick's work in a number of ways. For example, the mutation that produced telepathic twins Edie and Bill in *Dr Bloodmoney* is the result of nuclear fallout; in *Flow My Tears, the Policeman Said*, Alys and Felix Buckman are both members of a genetically engineered 'elite' class of posthumans; and two of the androids in *Do Androids Dream of Electric Sheep?* (1968) are so life-like as to render all the existing biometric tests for discerning humans from robots obsolete are described as twins. While twins constitute a recurring motif in the fiction of the 1960s and 1970s, they become particularly prominent in the work produced during the final years of Dick's life. This phase of Dick's output performs the characteristics that Bruce Clarke and Mark Hansen ascribe to second-order or what they call neocybernetics: 'the recursive complexities of observation, mediation, and communication'.[106] The period that Jameson designates as the 'religious' cycle, that includes both the *Valis* trilogy (1981–2) and the voluminous journal entries and theological mediations referred to as the *Exegesis* (which were first published in 2011) are characterized by the obsessive examination of an event that took place in early 1974. In the wake of a series of misfortunes, including having his house repossessed, a mental breakdown and separation from his fourth wife, Dick experienced a series of hallucinatory mystical visions. In later accounts of these visions, he would describe being visited by a divine or alien intelligence, which he labelled VALIS, an acronym for Vast Active Living Intelligence System. In the descriptions of this system developed in the novel *Valis*, which combines fictionalized autobiography with fragments of the science fictional gnosis published in full in the *Exegesis*, the influence of cybernetics is clear. The mystical entity that is revealed to him in his visions is described as the true status of the universe as 'living information'.[107] Entry number 48 from the *Exegesis*, reproduced in *Valis*, describes the universe as a computer, within which humans are stored memory: 'we appear to be memory coils (DNA carriers capable of experience) in a computer-like thinking system which, although we have correctly recorded and stored thousands of years of experiential information, and each of us possesses somewhat different deposits from all the other life forms, there is a malfunction – a failure – of memory retrieval'.[108] Salvation can only be achieved through a gnostic recuperation of this lost memory, a process he describes using the Greek term 'anamnesis (loss of amnesia)'.[109]

The traumatic process of overcoming this disjuncture between individual memory and the 'thousands of years of experiential information' stored on the universal computer is narrated using the tropes of twins and doubles. One of the effects of the mystical events of February and March 1974 (which Dick referred to as '2-3-74'), to use Dick's preferred designation for his breakdown, was to experience his psyche as being occupied by that of an early Christian called Thomas who lived during the first century AD. The California of 1974 became overlaid by ancient Rome in an effect that Dick describes as 'double exposure'.[110] He takes this experience of possession as evidence that 'we are not individuals' but rather 'stations in a single Mind'.[111] While 'we are supposed to remain separate from one another at all times', Dick's vision opened the doors to this transindividual level of experience, which is staged as a series of doublings and twinnings, a process indicated by the name of his early Christian occupant.[112] The name Thomas, which means 'twin', is a reference to early 'Thomasine' sects that believed that Christ's disciple Thomas was in fact his twin brother. Twins and doubles proliferate through the book. Dick creates a critical distance from his mystical experiences by ascribing them to a doppelgänger called Horselover Fat, a name that combines a literary translation of the Greet origins of the name Philip ('philippos' meaning 'friend of horses') and the translation of the German word 'dick' (meaning 'thick' or 'fat'). Connecting all these doubles together is the mythic figure of the absent twin. As Davis puts it, 'soaring angelic over all these secondary personalities and otherworldly cross-banding plasmates [a term used for informational organisms] is the absent-presence of Dick's own twin sister Jane'.[113] One of the many schizoid-Socratic dialogues in the *Exegesis* ends with the question: 'do you have any intuition or guess as to who and what the Valis mind is?' To which the answer is: 'Yes. It is female. It is on the other side – the postmortem world. It has been with me all my life. It is my twin sister Jane'.[114]

A mythological section titled 'TWO SOURCE COSMOGONY' describes the origins of our world as the result of the interaction between two cosmic twins. While both are the children of 'the One' and are 'motivated by a desire to *be*' (emphasis mine), one breaks out of its sac prematurely while the other only emerges at full term. The two develop through 'dialectical interaction' so that the premature twin (Form II) would be a 'hologram-like interface' that would serve as a 'teaching instrument' for true existence in the world, which is embodied by the full-term twin (Form I).[115] While the Form I twin is described as 'information-rich', the Form II twin is characterized by 'noise' and 'entropy', the effect of which is to impair its 'teaching function' and lead it down a path of decay and disorder. Having failed to save the Form II universe by sending an emissary ('Christ'), the Form I universe eventually kills its 'deranged twin' which continues to live a kind of illusory half-life.[116] This account of the birth of the informational universe splices together narratives of gnostic heavenly twins with the cybernetic concept that equates life with information and negative entropy. Rather than forming a closed system, the relationship between the twins evokes that of Maxwell's Demon, the thought experiment created by mathematician James Clerk Maxwell in which the second law of thermodynamics is violated. According to the experiment, the 'demon' that controls the door between two chambers quickly opens and shuts it to let only the fast molecules into one chamber, which heats up (increasing entropy) while the other cools down (decreasing entropy). In his reading of Heinz von Foerster's 1959 paper 'On Self-Organising Systems and Their Environments', which he argues foreshadows second-order cybernetics' preoccupation with recursion and open systems, Bruce Clarke focuses on the use of Maxwell's Demon. In von Foerster's account, Maxwell's Demon does not appear as a 'solo act within a closed system, but instead as a duo of demons collaborating to self-organise within a system/environment dyad under their mutual observation'.[117] Dick's techgnostic take on this thought experiment echoes that of von Foerster in its narrative of 'collaborative' twin demons. He presents our world as trapped in noisy entropic decline but providing glimpses through a demonic door into an information-rich twin chamber.

In the final pages of *Valis*, the narrator, who has eventually revealed himself to be Philip K. Dick himself, describes his double Horselover Fat as a medium through which he can communicate telepathically with VALIS. 'A voice in my head spoke. *Tell her radiation bothers you.* It was the AI voice which Horselover Fat had heard since 1974; I recognised it.'[118] This passage contains a number of disjunctions, including a gap that opens up between the author and Horselover Fat as well as between Dick/ Horselover Fat and Valis. Once the author has acknowledged the fact that Horselover Fat was created as a narrative device, he starts to function more like a twin than a double: following a developmental path that is distinct from that of Philip K. Dick. And yet, they are both bound together by Valis, the information 'system' in which they are immersed and of which they form a part. They perform for each other the function of gateways: porous membranes onto Valis, both protection and opening, both inside and outside each other. The relationship between Dick, Fat and Valis is used as a metaphor for the cybernetic concept of creativity that infused the period of the *Exegesis*. Clarke and Hansen point out that the shift from first-order to second-order cybernetics 'marks the passage to a general form of recursivity that can spiral outwards and thereby create the new at successively higher levels'.[119]

Throughout the pages of the *Exegesis*, Dick repeatedly studies and works through his own fictional writing from earlier in his life. Novels such as *Ubik* and *Flow My Tears, the Policeman Said* assumed the status of holy scripture that, if scrutinized with enough care and attention, might reveal crucial gnostic insights about the true nature of the information-universe. In other words, his own writing started to occupy an ambiguous space of internal externality. As Davis explains, within the *Exegesis*, 'Dick is generally loathe to make his person – rather than his texts – the locus of messianic power. More typically, he casts himself as a more or less passive relay node in a salvational network, an ignorant messenger who channels texts that know more – and do more – than he does.'[120] The absent-presence of the twin provides a model for the cybernetic relay of information. The twin framework, therefore, is a key aspect of the conceptual impetus for the shift that James Burton argues is both 'registered' and 'enacted' by the *Exegesis*: 'a shift from a hermeneutic to a post-hermeneutic understanding of "code" [...] in twisting the (exegetical) search for meaning into the production of dynamic, informational forms'.[121] A shift from exegesis to the 'ecologization of thinking' announced by Hörl.

Conclusion: Cybernetic ontologies of the enemy

Alongside this shift to an ecological understanding of human subjects in processes of 'dynamic' information production, the mythotechnesis of twins in Philip K. Dick's fictional world also serves to perform the relationship between self and enemy that was foundational to cybernetic theory. As discussed in relation to *More Than Human*, Norbert Wiener developed the central principles of cybernetics while attempting to devise a mathematical mechanism through which ground-based artillery could hit fast-moving bombers with a greater degree of accuracy. Peter Galison argues that cybernetics was predicated on 'a picture of a particular kind of enemy [that] was so merged with machinery that (his) human-nonhuman status was blurred'.[122] Furthermore, in the process of fighting this cybernetic enemy, 'Wiener and his team began to conceive of the Allied antiaircraft operators as resembling the foe, and it was a short step from this elision of the human and the nonhuman in the ally to a blurring of the human-machine boundary in general'.[123] The cybernetic vision was developed through two processes of duplication. Firstly, the antiaircraft predictor conceptualized by Wiener's team would simulate the most likely flight path that a bomber would take and feed this forward to the aircraft gun. Secondly, by constructing a vision of the enemy as a human-nonhuman assemblage, the cybernetics theorists produced a version of the allies, and the self in general, in its mirror image. Cybernetics is premised on a doubling between self and other. In contrast to the conception of otherness that subtended the period of European Imperial domination (articulated most clearly through Edward Said's concept of Orientalism) the cybernetic enemy is more like Wiener's description of the opposing players in Von Neumann's game theory: 'perfectly intelligent, perfectly ruthless operators'. As Galison puts it, the cybernetic enemy is a 'theoretical representation in which information, statistics, and strategies are applied to moves and countermoves in a world of opposing but fundamentally like forces'.[124] From this perspective, rather than the outcome of unconsummatable desire, the flickering jig that takes place between Grey Walter's robot tortoises could be seen as a death dance of mutual destruction. Elmer and Elsie are enemies constructed in each other's likeness, doomed to endlessly mirror the other's moves through the projection of their own programmed proclivities.

This cybernetic conception of otherness is played out in Dick's 1968 novel *Do Androids Dream of Electric Sheep?* When the 'bounty hunter' Rick Deckard sets out to 'retire' the final three escaped members of the latest, super-intelligent series of Nexus-6 androids, he calls on the help of another Nexus-6 android, a prototype used by the Rosen corporation as a display model to sell its new products. When the prototype android, Rachel Rosen, reads the profile of the remaining androids from Deckard's reports, she is shocked to discover that one of them, who calls herself Pris Stratton, 'is the same type as I am'.[125] 'Didn't you notice the description?' she asks Deckard. 'It's of me, too. She may wear her hair differently and dress differently [...]. But when you see her you'll know what I mean.'[126] Attempting to explain her emotional reaction, which comes as a surprise to Rachel, she expresses an ambivalence. At first, she interprets her reaction as empathy, which in the world view of the bounty hunters and the state apparatus they represent is the *sine qua non* of humanity. Androids are only proved as such if they fail the Voigt-Kampff empathy test. 'You know what I have? Toward this Pris android? [...] Identification; there goes I.'[127] However, she goes on to interpret the doubling between her and Pris as evidence of her lack of humanity. Her android twin is a reminder that 'we *are* machines, stamped out like bottle caps'.[128] She fears that she will be replaced by Pris and killed in her stead. At this point, when Deckard draws a comparison between Rachel-Pris and 'human identical twins', Rachel is quick to point out the difference: 'But they identify with each other; I understand they have an empathic, special bond'.[129]

Mirroring takes place in a number of ways in this sequence. Firstly, the logic behind asking Rachel to help Deckard track down the final three androids is that she can pre-empt their moves: 'I know Nexus-6 psychology. That's why I'm here'.[130] In the battle between bounty hunter and escaped androids, Rachel is used in a way that echoes Wiener's antiaircraft predictor. Rachel also compares her reaction to Pris to the role played by empathy in dominant official descriptions of human subjectivity, an emotional bond for which identical twins constitute the most complete embodiment. In other words, Rachel mirrors herself against dominant conceptions of the human. Implicit in the scene is a further mirroring dynamic, between Deckard and Pris. The main narrative tension in the novel is provided by the growing empathy that Deckard feels for androids, an emotion that he suspects might be incompatible with the profession of bounty hunter. The plea to Rachel for help was partly a pretext for sleeping with her, an activity that Deckard hopes might cure him once and for all of his android empathy.

However, far from diminishing his empathy, Rachel's ambivalent reaction to Pris is deeply familiar to Deckard. Rather than help him reinforce distinctions between humans and machines, the display of android twinship performed by Rachel serves to further blur the boundaries between the two and sets the stage for the vision of ecological entanglement with which the novel ends, in which the human is inextricable from the technological environment. Twinship in the novel is used as a framework for relationships between self and other that lead not to the policing of boundaries between human and non-human, life and non-life, but to an acceptance of entanglements and the non-modern ontologies that they enable. In Dick's vision, twins are caught in a dance of recursion that constitutes a motor of creativity and the production of the new.

Notes

1. Vladimir Nabokov, 'Scenes from the Life of a Double Monster', in *Nabokov's Dozen: Thirteen Stories* (London: Penguin Books, 1971 [1958]), 137.
2. Nabokov, 'Scenes from the Life of a Double Monster', 136–41.
3. Nabokov, 'Scenes from the Life of a Double Monster', 137.
4. William F. Barrett et al., 'Report of the Literary Committee', *Proceedings of the Society for Psychical Research* 1 (1882–3): 147.
5. Frederic W.H. Myers, *Human Personality and Its Survival of Bodily Death, Volume I* (London: Longmans, Green, and Co., 1903), 427.
6. Guy Lyon Playfair, *Twin Telepathy* (Guildford: White Crow Books, 2012), 5.
7. Frederic W.H. Myers, 'The Subliminal Consciousness. Chapter 1: General Characteristic and Subliminal Messages', *Proceedings of the Society for Psychical Research* 7 (1892): 305.
8. Nabokov, 'Scenes from the Life of a Double Monster', 135–6.
9. See Ellen Pifer, 'Locating the Monster in Nabokov's "Scenes from the Life of a Double Monster"', *Studies in American Fiction* 9:1 (1981): 97–101.
10. N. Katherine Hayles, *How We Became Posthuman: Virtual Bodies in Cybernetics, Literature, and Informatics* (Chicago and London: The University of Chicago Press, 1999), 2.
11. Norbert Wiener, *The Human Use of Human Beings: Cybernetics and Society* (London: Free Association Books, 1989 [1950]), 96.
12. Ute Holl, 'Trance Techniques, Cinema and Cybernetics', in *Trance Mediums and New Media: Spirit Possession in the Age of Technical Reproduction*, ed. Heiki Behrend, Anja Dreschke, and Martin Zillinger (New York: Fordham Scholarship Online: 2015), 266.

13 Félix Guattari, *The Three Ecologies*, trans. Ian Pindar and Paul Sutton (London and New Brunswick, NJ: The Athlone Press, 2000 [1989]), 36.
14 Erich Hörl, 'Introduction to General Ecology: The Ecologization of Thinking', in *General Ecology: The New Ecological Paradigm*, ed. Erich Hörl with James Burton and trans. Nils F. Schott (London: Bloomsbury, 2017), 6–8.
15 Hörl, 'Introduction to General Ecology', 9.
16 Hörl, 'Introduction to General Ecology', 8.
17 Nabokov, 'Scenes from the Life of a Double Monster', 141.
18 John W. Campbell in a letter to Eric Frank Russell, quoted in Damien Broderick, *Psience Fiction: The Paranormal in Science Fiction Literature* (Jefferson, NC: McFarland & Company, 2018), 14.
19 David Burrows and Simon O'Sullivan, *Fictioning: The Myth-Functions of Contemporary Art and Philosophy* (Edinburgh: Edinburgh University Press, 2017), 342.
20 Roger Luckhurst, *The Invention of Telepathy 1870–1901* (Oxford: Oxford University Press, 2007), 60.
21 Guattari argues that '[n]ow more than ever, nature cannot be separated from culture; in order to comprehend the interactions between eco-systems, the mecanosphere and the social and individual Universes of reference, we must learn to think "transversally". *The Three Ecologies*, trans. Ian Pindar and Paul Sutton (London and New Brunswick, NJ: The Athlone Press, 2000 [1989]), 43.
22 Hayles, *How We Became Posthuman*, 63.
23 W. Grey Walter, *The Living Brain* (London: Penguin Books, 1961 [1953]), 115.
24 Walter, *The Living Brain*, 116.
25 Andrew Pickering, *The Cybernetic Brain: Sketches of Another Future* (Chicago and London: University of Chicago Press, 2010), 21.
26 Norbert Wiener, *The Human Use of Human Beings: Cybernetics and Society* (London: Free Association Books, 1989 [1950]), 101–2.
27 Wiener, *The Human Use of Human Beings*, 102.
28 Wiener, *The Human Use of Human Beings*, 102.
29 Wiener, *The Human Use of Human Beings*, 102.
30 This use of twins is also echoed in a mention of twins in *The Living Brain* that is not directly connected to Elmer and Elsie. In a chapter on how the brain wave patterns provided by electroencephalographic (EEG) records are as unique to individuals as fingerprints, Walter points out that '[n]ot even identical or uniovular twins have quite identical alpha patterns.' Furthermore, the development of these 'brainprints' over the course of a lifetime are conditioned by experience. These acquired differences in brain patterns are best studied in monozygotic twins. 'The similarity of their brain mechanisms may continue to be as close as their physical resemblance when they mature, but their conditioning experience will not have been precisely alike'. Walter, *The Living Brain*, 172 and 188.

31 Walter, *The Living Brain*, 103.
32 Walter, *The Living Brain*, 103.
33 Lily E. Kay, 'Cybernetics, Information, Life: The Emergence of Scriptural Representations of Heredity', *Configurations* 5:1 (1997), 24.
34 Kay, 'Cybernetics, Information, Life', 30.
35 Quoted in Kay, 'Cybernetics, Information, Life', 26.
36 Hayles, *How We Became Posthuman*, 5.
37 Hayles, *How We Became Posthuman*, 14.
38 Pausanias, *Description of Greece, Volume IV: Books 8.22–10* (Arcadia, Boetia, Phocis and Ozolian Locri), trans. W. H. S. Jones. Loeb Classical Library 297 (Cambridge, MA: Harvard University Press, 1935), 311.
39 A.E. van Vogt, *Slan* (London: Panther Books, 1960 [1940]), 68.
40 Damien Broderick, *Psience Fiction: The Paranormal in Science Fiction Literature* (Jefferson, NC: McFarland & Company, 2018), 18.
41 Heiki Behrend and Martin Zillinger, 'Introduction: Trance Mediums and New Media', in *Trance Mediums and New Media: Spirit Possession in the Age of Technical Reproduction*, ed. Heiki Behrend, Anja Dreschke, and Martin Zillinger (Fordham: Scholarship Online, 2015), 4.
42 See Jeffrey Sconce, *Haunted Media: Electronic Presence from Telegraphy to Television* (Durham, NC: Duke University Press, 2000).
43 Robert A. Heinlein, *Time for the Stars* (Rockville, MD: ARC Manor, 2017 [1956]), 12.
44 Heinlein, *Time for the Stars*, 13.
45 Heinlein, *Time for the Stars*, 108–9.
46 Heinlein, *Time for the Stars*, 109.
47 Heinlein, *Time for the Stars*, 113.
48 Heinlein, *Time for the Stars*, 12.
49 Heinlein, *Time for the Stars*, 184.
50 Luckhurst, *The Invention of Telepathy 1870–1901*, 183.
51 A. E. van Vogt, *The Weapon Makers* (Los Angeles: Agency Editions, 2013 [1943]), 3.
52 van Vogt, *The Weapon Makers*, 9.
53 Fredric Jameson, 'The Space of Science Fiction: Narrative in van Vogt', in *Archaeologies of the Future: The Desire Called Utopia and Other Science Fictions*, ed. Fredric Jameson (London: Verso, 2007), 316.
54 Jameson, 'The Space of Science Fiction', 50.
55 Jameson, 'The Space of Science Fiction', 2.
56 Jameson, 'The Space of Science Fiction', 2.
57 A.E. van Vogt, *Slan* (London: Panther Books, 1960 [1940]), 134.
58 van Vogt, *The Weapon Makers*, 71.
59 van Vogt, *The Weapon Makers*, 167.
60 van Vogt, *The Weapon Makers*, 165.

61 van Vogt, *The Weapon Makers*, 164.
62 van Vogt, *The Weapon Makers*, 46.
63 van Vogt, *The Weapon Makers*, 132.
64 In a typical narrative cited by Playfair, a father of twin daughters from Rockland, Maine, accidentally slams the car door on one of his daughter's hands. 'She of course yelped right away [...]. Her sister yelped too. [...]. That's right – Twin A slammed her fingers in the car door and Twin B got the crease and bruises.' Playfair, *Twin Telepathy*, 143.
65 Playfair, *Twin Telepathy*, 110.
66 Theodore Sturgeon, *More than Human* (London: Orion, 2000 [1953]), 194.
67 Sturgeon, *More than Human*, 196.
68 Peter Galison, 'The Ontology of the Enemy: Norbert Wienr and the Cybernetic Vision', *Critical Inquiry* 21:1 (1994): 232.
69 Doug Moench and Alex Niño, *Heavy Metal Presents Theodore Sturgeon's More Than Human: The Graphic Story Version* (New York: Byron Press Visual Publications, 1978), 85.
70 Moench and Niño, *Heavy Metal Presents Theodore Sturgeon's More Than Human*, 85.
71 Moench and Niño, *Heavy Metal Presents Theodore Sturgeon's More Than Human*, 76.
72 Moench and Niño, *Heavy Metal Presents Theodore Sturgeon's More Than Human*, 72.
73 Guattari, *The Three Ecologies*, 34.
74 Guattari, *The Three Ecologies*, 34 and 45.
75 Sturgeon, *More Than Human*, 5–6.
76 Moench and Niño, *Heavy Metal Presents Theodore Sturgeon's More Than Human*, 27.
77 Moench and Niño, *Heavy Metal Presents Theodore Sturgeon's More Than Human*, 27.
78 Maurizio Lazzarato, *Signs and Machines: Capitalism and the Production of Subjectivity*, trans. Joshua David Jordan (Los Angeles: Semiotext(e), 2014), 13.
79 Lazzarato, *Signs and Machines*, 12.
80 Lazzarato, *Signs and Machines*, 25.
81 Lazzarato, *Signs and Machines*, 27.
82 Lazzarato, *Signs and Machines*, 49.
83 Moench and Niño, *Heavy Metal Presents Theodore Sturgeon's More Than Human*, 48.
84 Charles Hatfield, 'An Art of Tensions', in *A Comics Studies Reader*, ed. Jeet Heer and Kent Worcester (Jackson: University Press of Mississippi, 2009), 132.
85 Quoted in Lawrence Sutin, *Divine Invasions: A Life of Philip K. Dick* (New York: Carroll & Graf Publishers, 2005), 25.
86 Sutin, *Divine Invasions*, 12.

87   Roger Luckhurst, 'Diagnosing Dick', in *The World According to Philip K. Dick*, ed. Alexander Dunst and Stefan Schlensag (New York: Palgrave Macmillan, 2015), 19. Quoting from Ian Hacking, 'Making up People', in *Reconstructing Individualism: Autonomy, Individuality, and the Self in Western Thought*, ed. Thomas C. Heller, Morton Sosna and David E. Wellbery (Stanford, CA: Stanford University Press, 1986), 222–36.
88   Luckhurst, 'Diagnosing Dick', 16.
89   Luckhurst, 'Diagnosing Dick', 16.
90   Erik Davis, 'High Weirdness: Visionary Experience in the Seventies Counterculture', Diss. Rice University, Houston, Texas, 2015, 459–77.
91   Davis, *High Weirdness*, 477.
92   Davis, *High Weirdness*, 477.
93   Hayles, *How We Became Posthuman*, 188.
94   Fredric Jameson, 'History and Salvation in Philip K. Dick', in Fredric Jameson, *Archaeologies of the Future: The Desire Called Utopia and Other Science Fictions* (London: Verso, 2007), 363. Jameson proposes dividing Dick's oeuvre into three cycles: 'the so-called mainstream novels, from 1955 to 1960 (some seven novels that we still have); the Science Fiction period, from 1961 to 1968 (I include ten novels […]); and finally the religious novels, from 1973 to 1981 (some five works).' Jameson, 'History and Salvation in Philip K. Dick', 363.
95   Jameson, 'History and Salvation in Philip K. Dick', 366.
96   Fredric Jameson, 'After Armageddon: Character Systems in Dr Bloodmoney', in *Archaeologies of the Future: The Desire Called Utopia and Other Science Fictions*, ed. Fredric Jameson (London: Verso, 2007), 358.
97   Hayles, *How We Became Posthuman*, 166.
98   Hayles, *How We Became Posthuman*, 166.
99   Hayles, *How We Became Posthuman*, 166.
100 Hayles, *How We Became Posthuman*, 166.
101 Jameson, 'After Armageddon', 362.
102 Hayles, *How We Became Posthuman*, 182.
103 D.W. Winnicott, *Playing and Reality* (London: Routledge, 1991 [1971]), 5.
104 Winnicott, *Playing and Reality*, 146.
105 Tania Espinoza, 'The Technical Object of Psychoanalysis', in *Stiegler and Technics*, ed. Christina Howells and Gerald Moore (Edinburgh: Edinburgh University Press, 2013), 153.
106 Bruce Clarke and Mark B. N. Hansen, 'Introduction: Neocybernetic Emergence', in *Emergence and Embodiment: New Essays on Second-Order Systems Theory*, ed. Bruce Clarke and Mark B. N. Hansen (Durham and London: Duke University Press, 2009), 6.
107 Philip K. Dick, *Valis* (London: Gollancz, 2001 [1981]), 77.

108 Dick, *Valis*, 108.
109 Dick, *Valis*, 108.
110 Dick, *Valis*, 124.
111 Dick, *Valis*, 124.
112 Dick, *Valis*, 124.
113 Davis, *High Weirdness*, 570.
114 Philip K. Dick, *The Exegesis of Philip K. Dick*, edited by Pamela Jackson and Jonathan Lethem (London: Gollancz, 2011), 520.
115 Philip K. Dick, *Valis* (London: Gollancz, 2001 [1981]), 103.
116 Dick, *Valis*, 104.
117 Bruce Clarke, 'Heinz von Foerster's Demons: The Emergence of Second-Order Systems Theory', in *Emergence and Embodiment*, eds. Clarke and Hansen 46.
118 Clarke, 'Heinz von Foerster's Demons', 227.
119 Clarke and Hansen, 'Introduction', 11.
120 Davis, *High Weirdness*, 573.
121 James Burton, 'From Exegesis to Ecology', in *The World According to Philip K. Dick*, ed. Alexander Dunst and Stefan Schlensag (New York: Palgrave Macmillan, 2015), 211.
122 Galison, 'The Ontology of the Enemy', 233.
123 Galison, 'The Ontology of the Enemy', 233
124 Galison, 'The Ontology of the Enemy', 264.
125 Philip K. Dick, *Do Androids Dream of Electric Sheep?* (London: Gollancz, 2007 [1968]), 163.
126 Dick, *Do Androids Dream of Electric Sheep?*, 163.
127 Dick, *Do Androids Dream of Electric Sheep?*, 164.
128 Dick, *Do Androids Dream of Electric Sheep?*, 164.
129 Dick, *Do Androids Dream of Electric Sheep?*, 164.
130 Dick, *Do Androids Dream of Electric Sheep?*, 164.

2

Twins as weird media

Twinship has long been a favoured trope of the horror genre. *The Shining*'s Grady twins, inspired by Diane Arbus's 1967 photograph 'Identical Twins, Roselle, New Jersey', has become one of the most iconic horror images of all time, the inspiration of countless parodies and Halloween costumes. In narratives ranging from nineteenth-century gothic tales to contemporary techno-horrors, twins and the bond that exists between them have been constructed as a form of living technology, points of mediation between either two halves of a split self, distinct ontological dimensions or the living and the dead. In a manner that echoes their role in social media, outlined in the Introduction, twins in horror occupy the ambiguous position in which connectivity and transparent communication coincide with their opposites: disconnection and the failure of communication. The long association between twins and the aesthetic effect of the uncanny goes a long way towards explaining the characterization of twinship as an excessive, unnatural connectivity. According to Freud's influential account, the feeling of the 'unheimlich' takes those who experience it back to a point in development of the psyche in which distinctions between the subject and the object world had not yet been formed and pervasive connectedness still prevails. It is, for Freud, 'a regression to a time when the ego had not yet marked itself off sharply from the external world and from other people'.[1] But twins have also been used to point to the limits of connectivity, the points at which communication fails. Conjoined twinship, for instance, is often presented as something unintelligible, a phenomenon that fundamentally ruptures our systems of making sense of selfhood and its connection with the world. Twins, in these cases, present conceptual blockages.

The ambiguities of twinship as a form of mediation in the horror genre – its oscillation between connection and disconnection – are played out in the 2016 supernatural thriller *Personal Shopper*, directed by Olivier Assayas. The film centres on the character of Maureen (played by Kristen Stewart) who is mourning her twin brother Louis following his death from a heart condition that is shared by both siblings. Like her brother, a convinced and dedicated spirit medium, Maureen is able to communicate with the dead and has vowed to stay in Paris, the place in which Louis died, until she has been told by him that she can move on with her life. The film opens with Maureen spending the night in Louis's old house in the suburbs, hoping to be contacted by her dead brother. While she is not acting as a spirit medium, Maureen works as a personal assistant and shopper for a high-profile model called Kyra. The plot of the film intertwines two narrative threads: Maureen's attempt to communicate with her brother and come to terms with his death, and Maureen's involvement in the investigation surrounding the brutal murder of Kyra. Cultural expectations about the two extremes of twin connection and disconnection condition the viewer's response to the action. Is the connection between Maureen and Louis so strong that it excludes the rest of the world and that death is no obstacle between them? Or did the traumatic and sudden absence of Louis trigger in Maureen a psychotic opening to the world, a stripping away of psychic defences, that culminated in murder?

Personal Shopper dedicates an unusual amount of screen time to other screens. The two plot strands converge when Maureen starts to receive a series of anonymous text messages on her mobile phone. While travelling by Eurostar to London to pick up a dress from a top designer, Maureen obsessively reads and responds to these messages. The viewer shares her distracted perspective as she absently negotiates the train stations and interacts with the sales assistant while being really focused on the buzz of her iPhone and the flashing dots of pending communication as the mysterious interlocutor types out a message. But while Maureen is initially convinced that the messages are written by her dead brother Louis from beyond the grave, the viewer is led to believe that they are being sent by Kyra's deeply suspicious jilted ex-lover – the man who ultimately admits to her murder. Rather than individual psychology or belief in the supernatural, the film is really about the type of contact and connection enabled by communications technologies. Maureen's ability to connect with her dead brother comes at the price of complete disconnection from the rest of the world. Far from an aberrant case of twin autism, Maureen is presented as just an extreme version of everyday communication in which digital connectivity seems to come at the price of a disconnection from place. And Maureen embodies both ends of the spectrum. Walking through King's Cross station while reading her mysterious messages, she is the archetypical smartphone zombie, whose social existence online is kept alive by a social death in the physical world. But, on the other hand, her constant use of digital media is presented as an extension of her role as a spirit medium.

This is made explicit in a sequence in which Maureen learns about the history of Spiritualism in the nineteenth century through a YouTube documentary that she watches through her iPhone. One of the professorial talking heads in the video points out that the Spiritualists who claimed to be able to communicate with the dead were technological pioneers, what you might nowadays call early adopters. The method of tapping out messages favoured in séances was first used not long after the invention of Morse Code, while the development of the emerging technologies of telephony and photography often held the promise not just to connect faraway places and preserve moments from the past but also to open a line of communication with the afterlife.[2]

In *Personal Shopper*, the bond that exists between twins is equated with the 'impossible' communication of spirit mediumship. In this way, the film is an example of how twinship is often associated with what Eugene Thacker has called 'dark media', a term he uses to describe the paradox of mediation as 'those moments when one communicates with or connects to that which is [...] inaccessible'.[3] In Claude Shannon's theory of communication, the act of communication presupposes the prior separateness of two entities. 'Point A' can only communicate with 'point B' if the two are distinct. '[T]he possibility of connecting points A and B relies on this notion of prior separation – the conditions of connection relying on a prior state of disconnection.'[4] A particularly dominant tendency of the contemporary horror genre explores mutations in Shannon's diagram of communication produced by the proliferation of media devices and the diffusion into everyday life of the act of communication by the constant production and distribution of information. In films of haunted or monstrous communication devices – from the fleshly telephone in Wes Craven's *A Nightmare on Elm Street* (1985) to the haunted TV sets of Hideo Nakata's paradigmatic J horror of 1998, *Ring* – 'media shift from the connection of two points in a single reality, to an enigmatic and ambivalent connection with an unnamed "beyond"'.[5] In this horror imaginary, rather than 'render the inaccessible accessible', the function of media is to 'reveal inaccessibility in and of itself'.[6] Twins in mass culture often embody 'enigmatic and ambivalent' connections. As we will see, they are often caught between dimensions or linger in limbo between separate realities. In *Personal Shopper*, twinship is constructed as a communications technology that, in a similar manner to J horror's haunted TV sets, is not malfunctioning but rather works too well. Twinship in *Personal Shopper* holds the promise of perfect communication. But perfect communication is an impossibility. The merging of points A and B ironically undermines the separateness that provides the conditions for communication in the first place. The bond between twins is used to expose the fact that connection and disconnection are one and the same thing.

The interplay between connection and disconnection in horror's characterization of twinship is evoked through the interaction between two aesthetic modes: the uncanny and the weird. If twins, as living 'doubles', have most frequently been associated with the aesthetic of the uncanny, in the first decades of the twenty-first century, during the consolidation of networked digital culture and the mounting urgency of the global climate catastrophe, twinship has become increasingly associated with weird aesthetics. Both uncanny and weird twins have been used to perform and question the ideologies of communications media and the forms of connectivity they enable. I start by exploring how twinship is constructed as an uncanny and weird form of mediation through an analysis of the role played by twins in the work of Edgar Allen Poe and H.P. Lovecraft as well as the 'new weird' fiction of Jeff Vandermeer. In the following section, I look at how eugenics discourse constructs twin bodies as a form of 'haunted' mediation that opens a window into the agential role of genetic inheritance on identity. Contemporary horror films such as *The Unborn* represent twinship as a form of mediation that is much more opaque – more weird – than its use in the popularization of behavioural genetics research would imply. I then discuss the trope of melancholy twins as a form of weird mediation in Fumiko Enchi's novel *Masks* (1958) and the 2019 documentary *Tell Me Who I Am*. Finally, I explore the association between twinship as weird mediation and the aesthetics of melancholy in Peter Greenaway's 1985 film *A Zed and Two Noughts*, and the light this sheds on how twins are used to blur cultural distinctions between technological and organic forms of mediation.

Uncanny vs. weird twins

The points of divergence between uncanny and weird uses of twinship is clear from the differences between two short stories frequently taken to be paradigmatic of the narrative of the two aesthetic modes: Edgar Allan Poe's 'The Fall of the House of Usher', published in 1839, and H.P. Lovecraft's 'The Dunwich Horror', published nearly a century later in the April 1929 issue of *Weird Tales*. The north American horror writer H.P. Lovecraft, who published prolifically and almost exclusively in pulp magazines between 1917 and his death in 1937, has, over the past ten years or so, become canonized as the grandfather of an aesthetic said to characterize the current era of pervasive technologies and global climate catastrophe more than any other. 'The Fall of the House of Usher', on the other hand, enacts the associations between twinship and the gothic trope of the uncanny double that was to be so influential on the cinematic and literary horror twins of the twentieth and twenty-first centuries. The story is narrated from the perspective of an unnamed protagonist responding to a cry for help from his estranged childhood friend Roderick Usher. Arriving at Usher's dilapidated country estate, the narrator finds his old friend in a state of 'excessive nervous agitation' manifested through an acute 'inconsistency' of manner; sullen and sluggish one moment, highly agitated the next.[7] Furthermore, Roderick's sister Madeline is also slowly succumbing to an unnamed disease, the symptoms of which – 'a settled apathy' and 'gradual wasting away of the person' – are baffling her doctors. The troubling nature of the scene encountered by the narrator is compounded by the fact that the material structure of the family house itself, as well as the surrounding countryside, are in a similar state of decadence. Indeed, the physical house resembles its inhabitants to such a degree that locals use the term 'House of Usher' to refer collectively to both. An atmospheric contagion pervades the scene in which the borders between the human and material worlds – which the narrator usually experiences as reliable solid – have become fragile and porous.

Not long after his arrival, the narrator is told that Madeline has been overcome by her mysterious illness and that her corpse has been placed in a former dungeon in the basement of the house (beneath the narrator's bedroom) and is awaiting a proper burial at a later date. When Roderick takes the narrator to visit Madeline's body, he learns for the first time that the two of them were twins. The experience of inspecting the face of the coffin's 'tenant' is narrated in a way that concisely reveals how twinship is constructed in the story as a whole:

> A striking similitude between the brother and sister now first arrested my attention; and Usher, divining perhaps, my thoughts, murmured out some few words from which I learned that the deceased and himself had been twins, and that sympathies of a scarcely intelligible nature had always existed between them.[8]

Firstly, despite the fact that they are fraternal twins, it is the similarity between the siblings that is most immediately striking. Twinship in the story is used in a structurally very similar way to the trope of the gothic double that Poe's tales rejuvenated and helped to popularize. Roderick and Madeline are characterized by the symbiotic relationship that exists between them, hinted at by the fact that they share 'sympathies of a scarcely intelligible nature'. The climax of the story uncovers the extraordinary degree of their connection with one another. When Roderick reveals that his sister was buried alive, in a moment of unbearable horror, Madeline abruptly appears at the chamber door. As she falls into the room, Roderick's heart stops and he dies as if they had swapped positions and crossed the borders between life and death in opposite directions. The denouement opens up the possibility that Madeline was an outer emanation of Roderick's own fascination with and fear of death: a projection of his own neuroses. In that respect, twinship is mobilized in the story in a way that echoes the use of the double in the story 'William Wilson', which was published in the same year. In 'William Wilson', the narrator's double is described as his twin. When it emerges that the narrator's school companion – who looks like him, sounds like him and shares his already repetitious name – was also born on the same day, he remarks: 'assuredly if we *had* been brothers we must have been twins'.[9] The ending carries out a similar reversal to that which takes place in 'The Fall of the House of Usher'. Pursued throughout his youth and early adulthood, the narrator finally shoots his doppelgänger in a duel only to discover that it is his own chest that has been pierced by the bullet.

Secondly, returning to the key passage in 'The Fall of the House of Usher', in the narrator's account of gazing at Madeline's pallid countenance, twinship is described in terms of mediation. While Roderick and his twin sister are characterized by the ease of communication that exists between them, the bond that they share – their connection – is inscrutable to others. 'Sympathies' exist between them, but these sympathies are 'scarcely intelligible'. Throughout the story, Roderick himself is described as a highly effective medium of communication. In the face-gazing scene, the narrator suspects that Roderick has 'divined' his thoughts. Following the events in the basement, the narrator finds Roderick 'gazing upon vacancy for long hours, in an attitude of the profoundest attention, as if listening to some imaginary sound'. Roderick's strained senses appear to be tuned into frequencies beyond the realms of usual human experience. One of the main symptoms of Roderick's malady is the 'morbid acuteness' of his senses, which renders the smell of flowers 'oppressive' and even the faintest of light 'torture'.[10] Due perhaps to the intensity of his connection with the world around him, Roderick directs his senses 'inwards', particularly onto his connection with his sister Madeline. In a way that foreshadows the association between twinship and psionics, at the climax of the story Roderick seems to throw open the door separating him from his twin with the force of his words alone. The connection between Roderick and Madeline makes way for the more general sense of atmospheric contagion experienced by the narrator, resulting in a blurring of the boundaries between humans and the material world, life and death, the animate and inanimate. Meanwhile, the intensity of their twin connection comes at the expense of communication with the rest of the world. In the opening pages, the narrator alludes to rumours that the decline of the Usher family is the result of inbreeding and incest. '[T]he entire family lay in the direct line of descent, and had always, with very trifling and very temporary variation, so lain.'[11] The twinship of Roderick and Madeline both symbolizes and is presented as the end result of the Usher family's insularity. To return to Shannon's diagram of communication, the passage of information between the two siblings is so successful that it undermines the distinction between the two that made communication necessary in the first place. Twinship in 'The Fall of the House of Usher' takes mediation to its breaking point.

Poe's story, therefore, provides the blueprint for the association between twins and entropy that would be central to the trope of cybernetic twinship explored in Chapter 1. It has also been taken up by countless horror writers since Poe. In Bram Stoker's short story 'The Dualitists: Or the Death Doom of the Double Born', first published in 1886, the bond of twinship leads to violence and destruction. The narrative focuses on Harry and Tommy, boyhood friends who live on the same street and are so close that, '[c]ompared with these two youths, Castor and Pollux [...] are but tame examples of duality'.[12] In the house separating that of Harry and Tommy are born twins so 'alike in form, feature, size, expression, and dress [...] that one "might not have told either from which"'.[13] The twins are presented as the embodiment of Harry and Tommy's stifling proximity. When the boys are gifted two identical knives, they decide to test their quality 'by the ordeal of the Hack'.[14] This involves striking them against each other until one of them breaks. When both knives are damaged, they carry out the 'Hack' on their kitchen cutlery and then, when this runs out, they move onto other household objects and, eventually, animals. The horror of crushing live rabbits against each other only whets their appetite. Eager for a greater challenge, Harry and Tommy decide to carry out the Hack on their neighbours the twins. 'They are exactly equal! This is the very apotheosis of our art!'[15] When the father of the twins finds the boys on the roof of their house throwing them against each other, he tries to shoot them, misses and kills his children instead. The headless bodies of the twins then fall and kill their parents, making them 'posthumously guilty of the crime of parricide'.[16] The story ends with Harry and Tommy, now respectable grandparents, thinking back fondly on their youthful antics. It is as if killing the twins was the only way of separating the imaginary bond of twinship that existed between themselves. While the story playfully evokes a number of possible allegorical readings, it also takes to an absurd and gruesome extreme the gothic depiction of the bond of twinship as leading to decay.[17]

Many of these qualities were subsequently taken up by Freud in his attempts to account for and categorize the experience of the uncanny. Nicholas Royle describes the uncanny in terms that capture Poe's use of twinship. In a way that pre-empts Freud's account of the 'unheimlich', twinship is presented as the feeling of having 'a foreign body within oneself'; the 'experience of oneself as a foreign body'; or a 'sense of ourselves as double, split, at odds with ourselves'.[18] But 'The Fall of the House of Usher' also makes a connection between the uncanny and communications technology that is picked up on by Christopher Johnson in his essay 'Ambient Technologies, Uncanny Signs'. For Johnson, cybernetic technologies herald 'the most *uncanny* of moments in the recent history of technology'.[19] 'Because, with their capacity for automatic calculation and processing, these new technologies approximate, or promise to approximate, the processes of human intellection, they in turn bring to light the automatic and programmatic at the different levels of organization of the "human"'.[20] The foregrounding of twinship as a form of excessive mediation in 'The Fall of the House of Usher' is accompanied by a general sense of self-awareness of the forms of mediation involved in the genre of the gothic tale. Not only is the story intensely intertextual (the narrator lists a number of gothic works that he reads together with his host) but it also pre-empts the real effects the story can have on the affective world of the reader. In an attempt to calm Roderick's nerves, the narrator reads him a 'Medieval romance' and the description of sounds within the world of the story eerily coincide with noises occurring in the 'real world'. A 'most unusual screaming or grating sound' can be heard at the very moment that a dragon shrieks within the story. The splitting of the self evoked through the trope of twinship metaphorizes the multiple doublings produced by communications technologies and the mass media that was still in its infancy. Just as Roderick seems possessed by foreign spirits or animated by energies beyond his control, the tale seems to write itself through the cybernetic reorganization of already existing narrative elements. The text itself is uncanny since it reveals the 'automatic and programmatic' nature of its organization that the conventions of realism would have hidden from view.

Poe's use of twins and doubles to establish an association between communications technologies and the uncanny pre-empts its return as the dominant affective response to the automated technologies of the late twentieth and twenty-first centuries. The proliferation of doubles in the work of David

Lynch, for instance, is a testament to the influence of Poe's association of twinship and uncanny media. Although doubling is a key motif in all Lynch's work, *Lost Highway* contains his most overt riff on the theme of the gothic double.[21] The plot centres on the character of Fred Madison, a jazz musician who violently murders his wife in their house in the Hollywood Hills. In a dream-like sequence that takes place immediately prior to the murder, the musician (played by Bill Pullman) approaches an image of himself in a mirror hidden in a previously unknown corner of their bedroom. The second half of the film abruptly changes tack when Fred, following his incarceration for the murder, mysteriously and abruptly disappears leaving an entirely different person in his place: the car mechanic and petty criminal Pete. The splitting of Fred in the mirror paves the way for his replacement by Pete, in a way that echoes the gothic trope of inner violence being projected onto external doubles. It also sets the scene for a more generalized process of doubling as the two realities opened up by the split intersect and collide. The murdered wife Renee (played by Patricia Arquette) is doubled into the femme fatale gangster's moll Alice. In a key revelation sequence, the figure of twinship is evoked when Pete finds a photograph of brunette Renee and bottle-blonde Alice standing side by side. For a brief moment, which is soon undone by further twists of Lynch's serpentine story line, the idea that they are twins seems to offer the solution to the plot puzzle. Alongside the Freudian themes of narcissistic projection and paranoia, the trope of uncanny doubling is also associated with communications technologies. The viewer's response to the doubling of actors and the splitting of characters is conditioned by the film's focus on the fragmentation of subjectivity carried out by media technologies. Fred first learns of the violent death of his wife when he witnesses himself crouching over her corpse bathed in her blood. In one of the most memorable sequences in *Lost Highway*, the menacing 'Mystery Man' who has been haunting Fred's dreams approaches him in a party and tells him that, despite appearances, he is at his house. 'I'm there right now'. Fred calls his house on a mobile phone and hears the Mystery Man's voice, in a troubling dislocation of voice from presence. Lynch uses the trope of doubling to denaturalize everyday media devices and expose their disturbing uncanniness.

If Poe's 'The Fall of the House of Usher' uses twins in a way that exemplifies its association with the uncanny and particularly the uncanny nature of communications media, Lovecraft's story 'The Dunwich Horror' incorporates twinship into its 'weird' aesthetic and its exposure of the limits of communications technologies. The story centres of the character of Wilbur Whateley, born in the remote Massachusetts town of Dunwich in February 1913, to a 'sickly and pink-eyed' albino mother Lavinia Whateley and an unknown father.[22] A third-person omniscient narrator recounts the strange circumstances surrounding the Whateley household as Wilbur grows into a freakishly tall and 'goatish' scholar of 'rare and forbidden books' and his grandfather Old Whateley carries out a series of adaptations to the house that eventually leaves it as an empty shell. The dramatic denouement reveals that Wilbur was born with a twin brother, a monstrous representative of ancient malevolent forces set on reasserting their power on the contemporary world. Both Wilbur's research and his grandfather's home improvements were intended to set the stage for the evil twin to unleash his wrath upon the unsuspecting inhabitants of the Massachusetts town. Firstly, the treatment of twins in the story bears some close resemblances to its use by Poe, whose work was a key influence on Lovecraft. For instance, twinship in the story is associated with incest, insularity and racial degeneracy. One of the reasons why travellers in Massachusetts tend to steer clear of Dunwich, the reader is told, is its inhabitants' reputation for 'repellent decadence': 'They have come to form a race by themselves, with the well-defined mental and physical stigmata of degeneracy and inbreeding. The average of their intelligence is woefully low, whilst their annals reek of overt viciousness and of half-hidden murders, incests, and deeds of almost unnamable violence and perversity.'[23]

Secondly, twinship in the story brings together the two extremes of mediation. On the one hand, the twinship of Wilbur and his brother opens up a line of communication with the evil other-worldly forces referred to through the name 'Yog-Sothoth'. The twins work together as mediators for these ancient forces. Like Roderick in 'The Fall of the House of Usher', Wilbur is characterized as a highly efficient communicator. By the age of seven months, he had already become 'a fluent and incredibly intelligent talker.'[24] As a scholar, Wilbur is an effective networker, maintaining 'correspondence to many librarians in distant places',[25] and an able linguist, knowledgeable in a number of languages, including that of the dreaded tome of evil lore, the *Necronomicon*. While Wilbur is a canny manipulator of his chosen media technologies (books and letters), it is his twin brother's body that becomes a vessel of mediation as he embodies, in a fuller way than the rest of his family, the obscure power of Yog-Sothoth. However, the twinship of Wilbur and his brother also acts as a blockage to mediation, particularly visual mediation. The second half of the story focuses on the attempts of Henry Armitage, librarian of the fictional Miskatonic University, to stop the monstrous chain of events set in motion by the Whateley brothers. When Wilbur's twin starts to cause destruction along the valley surrounding Dunwich, Armitage and a few companions set out to kill him. They discover that he is invisible to human eyes and can only be rendered visible momentarily when covered in a magic powder. The narration of the twin's death is focalized through a group of Dunwich natives watching Armitage track the monster across a hillside through a 'pocket telescope of considerable power'. Even this most modern instrument of vision proves incapable of capturing the elusive twin and it is only when the powder is sprayed on the beast that they glimpse him in all his unimaginable horror. So twinship in 'The Dunwich Horror' constitutes both a magic form of mediation and a blockage to modern media.

However, while the construction of twinship bears key resemblances to 'The Fall of the House of Usher', there are differences between Poe's and Lovecraft's use of twins that mark the distinctions between uncanny and weird aesthetics. In Poe's tale, twinship is used to blur the boundaries between inside and outside the self, between life and death. In 'The Dunwich Horror,' by contrast, there is something inaccessible about the monstrousness represented by the twinship of the Whateley brothers. The forces that they mediate, the malevolent entity referred to as 'Yog-Sothoth', is characterized throughout the story by its inaccessibility. During Wilbur's visit to the Miskatonic University library, Armitage reads a translated passage of the *Necronomicon* over the goatish giant's shoulder containing a description of this evil power: 'The Old Ones were, the Old Ones are, and the Old Ones shall be. Not in the spaces we know, but *between* them, They walk serene and primal, undimensioned and to us unseen'.[26] One of the most resonant voices in calls for the canonization of Lovecraft as godfather of the weird has been that of philosopher Graham Harman. In his book *Weird Realism*, Harman sets out to make Lovecraft 'as great a hero to object-oriented thought as Hölderlin was to Heidegger'.[27] Lovecraft owes this status to his exploitation of two gaps in his writing: the first 'vertical' gap is between 'the real and the sensual'; the second 'horizontal' gap is between 'objects and their qualities'.[28] On the one hand, Lovecraft's tales repeatedly emphasize a disconnect between 'an ungraspable thing and the vaguely relevant descriptions that the narrator is able to attempt'.[29] Here, language falters in the face of the sheer otherness of the reality it confronts and consequently resorts to strategies of indirect allusion. On the other hand, when confronting this obstruction, rather than become 'enfeebled by an impossibly deep and distant reality', the language used in Lovecraft's descriptions of objects of horror often becomes 'overloaded by a gluttonous excess of surfaces and aspects of the thing'.[30] Rather than the allusiveness triggered by the 'vertical' gap, the 'horizontal' gap produces a 'cubist' form of writing, a quality that earned him the label of 'the twentieth century horror story's dark and baroque prince' (Stephen King).

Furthermore, Lovecraft repeatedly emphasizes these gaps by drawing attention to the medium of enunciation itself. In Lovecraft's world there is rarely direct access between the narration of events and the events being narrated. Rather, mediation is only ever partial, blocked by the failings of memory or the vagaries of hearsay and translation. Meanwhile, the expectations of the reader are frustrated by shifts in plot or perspective. Rather than anything to do with theme or content, Roger Luckhurst argues that the identifying characteristic of the weird is 'a waywardness that leaves the reader confounded at the slow mutation of the story out of one horizon of expectation and into another'.[31] While the economy of the uncanny is oddly reassuring – in that it 'always leads back to the ultimate familiar home [...] the womb' – the weird is characterized by a waywardness, a perpetual veering away from the familiar and expected.[32] 'The interpretive machinery of the uncanny inherently domesticates. In contrast, the monstrous breaches of the weird do not return us to something familiar but repressed, but instead veers away to invoke dread that is irreducible, that cannot be reductively interpreted, translated or returned'.[33] This process of 'veering' away from expectations is carried out both at the level of plot and the manipulation of genre. Weird fiction tends to play with distinctions between high and low culture, diverge from literary canons, celebrate obscure authors or invent imaginary ones.

While 'The Dunwich Horror' is an archetype of weird fiction as outlined by Luckhurst, twinship in the story is weird in both of the senses outlined by Harman. The description hesitates between allusion and cubism. For most of the story Wilbur's brother is hidden from view, his presence in the world alluded to by clues left for the knowing reader (the suspicious work done to the Whateley residence, for instance). When he is revealed, the description of him, like that of his brother, is a baroque montage of conflicting elements. As an overawed Dunwich resident puts it: 'It was an octopus, centipede, spider kind o' thing, but they was a haff-shaped man's face on top of it... '[34] Even when the hidden monstrous twin is revealed, the eye skates across its surface without gaining access to its essence, just as the viewer's gaze becomes caught in its multiple-articulated edges. Whereas Poe's uncanny twins hold the promise of access to dark truths lurking beneath the surface of things, Lovecraft's twins are all surface, or rather, a montage of surfaces. The first line of Otto Rank's *The Double* sums up the parallel he traces between psychoanalysis and the literary trope of the gothic double: 'The technique of psychoanalysis generally aims at uncovering deeply buried and significant psychic material, on occasion proceeding from the manifest surface evidence'.[35] By incorporating the Usher twins into this logic, they are rendered into an outer manifestation of a fundamental split at the heart of subjectivity which, in our everyday reality, remains hidden in plain sight. Lovecraft's weird twins, by contrast, merely give access to another dimensionless dimension that remains unknowable, and they therefore confront us with the limits of our media technologies and the knowledge we construct with them. Mark Fisher explains the difference between the weird and the uncanny in the following way: '[T]he weird is that *which does not belong*. The weird brings to the familiar something which ordinarily lies beyond it, and which cannot be reconciled with the "homely" (even as its negation). The form that is perhaps most appropriate to the weird is montage – the conjoining of *two or more things which do not belong together*'.[36] While the uncanny offers a glimpse into the depths of being, the weird offers nothing but a complex and dazzling articulation of surfaces.

A further 'weird' take on twinning is evident in Jeff Vandermeer's updating of Lovecraftian weirdness in The Southern Reach trilogy, published in 2014. The three books – *Annihilation, Authority* and *Acceptance* – revolve around the relationship between 'Area X', an area of wilderness in the United States that is being possessed by a mysterious unknowable force, and the 'Southern Reach', an equally mysterious agency set up to monitor it. One of the more startling effects of Area X is to absorb human visitors into its alien ecosystem and producing a clone that it sends back into the world beyond its borders. So, while *Annihilation* is narrated from the journals of the Biologist, a member of an all-female expedition into Area X and culminates in her apparent absorption, *Authority* recounts the 'return' to the Southern Reach agency of somebody who looks like the Biologist but asks to be called Ghostbird. The final instalment of the trilogy, *Acceptance*, focuses on the return of Ghostbird to Area X and uses twins and twinning as the main metaphor for describing the relationship between original and clone. The novel opens with the Director of the Southern Reach being absorbed into Area X: 'A kind of alien regard has twinned itself to you, easily mistaken for the atoms of the air if it did not seem somehow concentrated, purposeful. Joyful?'[37] This reference to twinning is echoed by Ghostbird's repeated references to the Biologist as her 'twin' and reinforced by descriptions of the two members of S&SB, an organization that is possibly responsible for the appearance of Area X, as 'deathly twins' whose experimental methods include 'necromantic doubling'.[38]

Much of the novel is taken up with Ghostbird's musings on the nature of her own existence and, in particular, her relationship with her 'twin', the Biologist. She describes having a highly mediated relationship with herself, an effect both of being a 'copy' and of the time she spent being interrogated by the Southern

Reach following her reappearance. She spends time 'examining the sense that her memories were not her own, that they came to her secondhand and that she could not be sure whether this was because of some experiment by the Southern Reach or an effect caused by Area X'.[39] She has a sense that she observes her memories as if 'through a window opening onto another person's life'.[40] At one point, Ghostbird speculates that Area X might have created her and the other doubles as a form of communication with the world beyond, a form of living media. But if this is the case, she muses, the message that she is being used to communicate is anything but clear. 'She might also be a message incarnate, a signal in the flesh, even if she hadn't figured out what story she was supposed to tell.'[41] Sara Wasson points out that the logic of the death drive is 'writ large in the majority of cloning fictions' which confront us with the 'dread possibility of repetition'.[42] On one level, *Acceptance* does follow this pattern with Ghostbird being drawn to her place of origin, a return to the womb. However, the relationship between Ghostbird and her 'twin', the Biologist, is not characterized by straightforward repetition but by mutation. This mutation operates on both sides of the 'split' and therefore undermines the distinction between original and copy.

When Ghostbird finally encounters the Biologist in Area X, the meeting is strongly reminiscent of the sighting of Wilbur Whateley's monstrous twin in 'The Dunwich Horror'. Ghostbird watches as she comes down the mountainside 'in all her glory and monstrosity [...] her body flickering and stitching its way into existence, in the midst of a shimmering wave that imposed itself on the reality of the forested hillside'.[43] The Biologist both seems to emanate from the landscape of Area X and occupy an uncertain liminal zone between distinct realities, which is occupied by Ghostbird, and another realm that remains inaccessible: 'that leviathan, still somehow half there and half not.... The edges wavery, blurred, sliding off into some *other place*'.[44] Here, we find both characteristics of weird realism outlined by Harman. Firstly, language seems to fail when confronted with an 'ungraspable thing'. At the heart of the Biologist lies a 'ponderous and muffled darkness'.[45] Secondly, the writing becomes overloaded by 'gluttonous excess': 'It had many, many glowing eyes that were also like flowers or sea anemones spread open, the blossoming of many eyes – normal, parietal, and simple – all across its body, a living constellation ripped from the night sky.'[46] And yet, although they remain fundamentally mutually inaccessible, peering at each other across an abyss, Ghostbird experiences the encounter as a 'communication or communion'.[47] It is this tension between mutation productive of new forms of life and a 'communion' with or embrace of otherness that is characteristic of weird twinning.

And yet the distinction between uncanny Poe and weird Lovecraft is not as clear-cut as perhaps this account of twinship in their work suggests. There is plenty of weirdness in Poe just as the uncanny exerts a powerful force on Lovecraft. In 'The Fall of the House of Usher', for instance, Roderick remains convinced of the 'sentience of all vegetable things' but that this sentience has a quality of 'inorganization' that renders it inaccessible to human senses. In his musings, he can but allude to the 'gap' that exists between the two. The opening description of Dunwich blurs the distinction between the natural and the artificial that echoes the uncanny obsession with automated animation. 'The summits are too rounded and symmetrical to give a sense of comfort and naturalness, and sometimes the sky silhouettes with especial clearness the queer circles of tall stone pillars with which most of them are crowned.'[48] It is useful here to explore the 'weirding' of Poe carried out by the Japanese horror and mystery writer of the 1920s–50s Taro Hirai in the association he makes between twins, the uncanny and technological mediation. In recognition of his literary idol, Hirai wrote under the penname of Edogawa Rampo, a phonetic transliteration of Edgar Allan Poe. The playful use of doubles and doppelgangers was a prominent feature of Rampo's career. His penname frames his writing as a mirror of Anglophone mystery writers. But if it is a mirror, then it is a distorted one, which twists narrative conventions consolidated by Poe to, as Patricia Welch puts it, 'tap into the elements of strangeness that Japan's headlong rush into modernity after the Meiji restoration had unleashed'.[49]

Furthermore, as a number of critics have pointed out, Rampo's oeuvre as a whole can be characterized by a divide between two genres: his scientific *honkaku* (orthodox) detective fictions and his *henkaku* (fantastic or weird) romances.[50] In practice, however, many of the stories play with readerly expectations to undermine distinctions between the orthodox and the weird and in the process unsettle the epistemological certainties of modernity. As well as a characteristic of Rampo's career and oeuvre, doubling is a common trope in his fiction in which it is often deployed to trouble the split between subject and object of the gaze that is constitutive of rational subjectivity. This is clearest in the story 'The Twins', which was first published in 1924 and, in a way that echoes Poe, associates twinship both with the trope of the uncanny double and technological mediation. The story is presented as a confession, made by a criminal charged with a murder perpetrated during a botched robbery, to a priest in which he confesses to a further crime that has thus far escaped the notice of the detectives: the murder of his twin brother. The motive for this initial crime was fraternal rivalry, both due to the elder twin's inherited wealth and his brother having married the woman he loved. '[T]he very reason for my wanting to kill him was that we were two persons in one. And how I *hated* my other half!'[51] But once dead, his 'other half' came back to haunt him to the point where he became afraid to look in the mirror for fear that it would be his brother looking back at him.

In the story, photography is used as a metaphor for the doubling of twinship. One day, the narrator explains, the younger twin found a fingerprint in one of his dead brother's notebooks. The sight alarms him because fingerprints, he realizes, are the one thing that differentiates the two brothers and could, therefore, expose the crime of identity theft. He decides to use the discovery to his own advantage. The younger twin takes a copy of the fingerprint and leaves it in the houses that he has started to burgle (to pay for his mounting debts) in the belief that their difference from his own prints will be the perfect alibi. It comes as a shock, therefore, when a detective comes to take his prints and finds that they are the perfect match for those found in the burgled house. It is only later, once the detective recounts a similar case, that the narrator realizes what happened. The fingerprint was his own. He didn't recognize it because the marks were produced by ink not from the ridges of his thumb print but the fissures between the ridges, 'producing a print like the negative of a photograph.'[52] It was only when the detectives took a photograph of the print and reversed the negative that they discovered the match with the prints of the accused. This plot twist triggers a series of associations between twinship and technological mediation that amplify those produced by Poe's story. Photographic negatives are used as a metaphor for the trope of the good/evil twin duality. This, in turn, points to connections that the story evokes between twinship and photography more generally. Both twins and photography have an ambiguous relationship with modern individualism and rational subjectivity. On the one hand, twins and photography share an uncanniness that unsettle the boundaries of individuality. The narrator sees himself externalized in his brother like somebody might feel themselves externalized in a photographic image.

On the other hand, both twins and photography are used to shore up modern individuality. If photography, due to its reproductive capabilities, unseats the uniqueness of the individual, it is also used to reinforce this uniqueness. The use of the photograph to connect the fingerprint to the criminal evokes the role of photography in the history of criminology traced by Allan Sekula.[53] Likewise, through the failure of the narrator's attempt at identity theft, the twins in the story seem to ultimately reaffirm distinctions between individuals. In this they would echo what Lisa Zunshine describes as the 'cognitively enjoyable' exercise offered by early modern twin plays such as Shakespeare's *Twelfth*

Night, discussed in the Introduction.[54] The 'essentialist biases' of spectators are challenged and then confirmed as the distinctions between twin siblings are blurred only to be ultimately reaffirmed.[55] Ultimately, however, despite its use of uncanny tropes, the story presents twinship as a form of weird mediation. According to the narrator, and in a way that echoes 'The Fall of the House of Usher', the surviving twin's body mediates the spirit of the dead twin. However, the only evidence the reader has of this is the word of the narrator himself since the crime of fratricide has gone undetected by the authorities and it is possible that the younger twin really has gone to work in Korea as he claimed. The reader has no way of knowing which twin is carrying out the confession. The reader does not know who is mediating who. The two are inextricable and the form of mediation they provide for one another is unclear. In this way, the story provides none of the consolations and reassurances of the uncanny and instead is structured around a constant alternation or 'veering' between perspectives, to use Luckhurst's term. Like the protagonists of Rampo's story, the twins that I trace throughout what remains of this chapter hesitate between the uncanny and the weird in their association of twinship with mediation. This ambiguity echoes the hesitation between molar and molecular that characterizes their use in contemporary behavioural genetics. But it is also indicative of the changing nature of communication as devices of digital connection become increasingly automated and embedded within the body.

Haunted vs. weird media

In his study of the treatment of media devices and processes of mediation in contemporary horror, Eugene Thacker makes a distinction between 'haunted' and 'weird' media. In both cases, the haunted or weird nature of the media devices in question does not result from the fact that they are broken, but rather that they are working 'at a level beyond that of traditional forms of human mediation'.[56] The difference is that, while haunted media forge connections between two distinct ontological orders (life and afterlife, for example) weird media emphasize a blockage or gap between these different realms. In the process, weird media draw attention to how, as Thacker puts it, 'all communication collapses back into a prior excommunication.'[57] Whereas haunted media are characterized by 'connection, communication, transparency', weird media produce 'disconnection, silence, opacity'.[58] These two forms of media map neatly onto the aesthetic modes outlined in the previous section. Haunted media are the technologies of the uncanny. In the hands of spiritualists of the nineteenth century, early photographic technologies became haunted when they opened lines of communication with the deceased. Their photographs produced a powerful effect of the uncanny blurring of the boundaries between life and death and bringing to the surface of the image the repressed and unruly affective dynamics of grief. Through his manipulation of horror conventions, by contrast, Lovecraft turns writing itself into a weird medium, capable only of gesturing to the abyss between realms through allusion and excess.

The trope of twinship is frequently used in narratives that borrow from the horror genre to exemplify both forms of media as well as explore the points of intersection between the two. In her 2009 novel *Her Fearful Symmetry*, Audrey Niffenegger establishes a connection between twinship and haunted media. The plot centres on two generations of 'identical' twins and exploits some of the most commonplace tropes of twinship, from the psychological claustrophobia of the twin bond to identity-swopping. Set in an apartment block overlooking Highgate Cemetery, the novel reworks gothic twin narrative devices within the conventions of mass-cultural literary realism.[59] Twinship in the novel renders the borders between life and death porous as both sets cross back and forth with varying degrees of success. In a nod to the fate of Madeline in 'The Fall of the House of Usher', one twin temporarily occupies a coffin within a family mausoleum before her corpse is revivified. Like Poe's story, *Her Fearful Symmetry* also makes a connection between the twin bond and writing as two forms of media capable of bestriding ontological divides and enduring after death. The connection is literalized when one of the protagonists communicates with the twin daughters of her own twin by tracing words on dusty surfaces using her ghostly finger. However, by asking its readers to firmly believe in its twin haunting, the book strips the gothic twin trope of its uncanny atmosphere. As outlined in the introduction, Olivier Assayas's film *Personal Shopper* also explores the narrative potential of twin mediumship. The film is rather more hesitant in its construction of twinship as a haunted medium than Niffenegger's novel. Assayas simultaneously maintains the plausibility of both supernatural and realist narrative explanations. Following one interpretation, the digital media devices clutched by the twin protagonist Maureen are haunted, enabling uncanny connection with the afterlife. Following the other interpretation, these ubiquitous communication devices are weird in that all they communicate is the impossibility of communication.

A more subtle example of slippage between twins being constructed as a form of haunted or weird mediation is provided by the 1958 novel *Masks*

(*Onna-men*) by the Japanese novelist Fumiko Enchi. The narrative centres on an elaborate act of revenge carried out by Mieko, the widowed wife of Matsugo, the head of the wealthy Togano family, who is mourning the death of her son Akio. The plot focuses on how Mieko manipulates the relationship between her daughter-in-law Yasuko, who has continued to live with her after the death of Akio, and two men who are in love with her. Central to Mieko's plotting is the revelation that Akio had a twin sister called Harume who was brought up in secrecy due to the family's shame at her learning difficulties and lingering folk beliefs that twinship is an aberration. Furthermore, the twins were fathered by her lover following a miscarriage that was induced by a maid with whom her husband was conducting a clandestine relationship. Using Yasuko as a lure, Mieko orchestrates a tryst between one of Yasuko's suitors and Harume, resulting in the latter's pregnancy. Although Mieko's motives are never rendered fully explicit, it is implied that her intentions for Harume's pregnancy were twofold. Firstly, the baby would bear the same blood as her beloved son Akio and therefore function as a living memorial to his life. Secondly, by perpetuating illegitimate strands of the Togano family, the baby would be an act of vengeance by polluting the purity of its bloodline.

The novel revolves around various forms of what Thacker terms 'haunted media'. Mieko, her daughter-in-law Yasuko and her two lovers share an interest in both Noh theatre and spirit possession. In an early scene, the four of them attend a séance in which the spirit of Akio speaks to his widow through a medium, who convulses 'as if filled with electricity' when possessed.[60] Following this sequence, Mieko and Yasuko visit the home of a Noh master in Kyoto to look at his famous collection of costumes and masks. When the master's assistant puts on one of the masks, Yasuko is overcome by the uncanniness of the spectacle. 'It was as if something dead had come to life [...] it was almost as if Akio's spirit had taken over the mask.'[61] At this, the narrator makes the following observation, focalized through Yasuko's suitor: 'Could it be that once again, as at the séance when the medium first spoke, she was caught up in the illusion that Akio had returned from the dead?'[62] The passage draws comparisons between Noh theatre and spirit possession as two forms of haunted media. As Doris Bargen puts it, in both, 'ordinary individuality is transformed as the actor or the possessed person is united with the aggrieved spirits of the dead or the living. The audience witness a liminal state of two personalities in one.'[63] These two forms of haunted mediation, and the undermining of 'ordinary individuality' that they enact, become a lens through which the reader is induced to view other relationships in the novel. This perspective in encouraged by one of Yasuko's lovers who, following the séance, observes that '[i]f Yasuko is the medium, then Mieko Togano is the spirit itself.'[64]

Both Noh masks and spirit possession are also used as metaphors for twinship in the novel. Yasuko is troubled by the likeness between Harume and her dead husband. 'Invest that face with wit and masculinity, and it would be, thought Yasuko, the face of Akio, so clearly had the bond of twinship been stamped in their looks.'[65] Yasuko likens this merging of the two faces to the uncanny feeling she had while watching the Noh master's male assistant put on the female mask. 'The moment when she had felt faint at the sight of the No mask on the Yakushiji stage, she had seen it plainly: the two faces of Harume and Akio coming together as one before her eyes.'[66] While this description is part of a longer connection between twin faces and masks explored in Chapter 5, in Enchi's novel the use of the mask serves to draw attention to how Harume becomes a medium for her dead brother. The fact that both the séance and the orchestration of Harume's affair are part of the process of mourning the death of Akio, principally for Mieko but also for Yasuko, is highlighted by the fact that one of the group members is a psychoanalyst with a side interest in folk tales about spirit possession. In this respect, in a way that echoes 'The Fall of the House of Usher', the bond of twinship is manipulated like a form of technological mediation with which to communicate with the dead.

And yet the form of mediation offered by twinship in the novel is ultimately more 'weird' than 'haunted' in Thacker's sense. Although not explicitly engaging with conventions of weird fiction, Enchi's novel shares several qualities with the narrative strategies described by Luckhurst. Enchi shares weird fiction's 'archive fever': 'hallucinating into existence phantoms of the catalogue that thumb their noses at the archons, the patriarchal authorities who guard the house of official documents'.[67] Much of the first section of *Masks* is taken up by a transcription of an essay authored by Mieko Togano and published in 1937 which, she later claims, had been written for the benefit of her lover. The essay proposes a new reading of the medieval classic of Japanese literature *The Tales of Genji* that places at its centre the relationship between the hero and the avenging spirit of his jilted lover and archetypal shamaness figure Lady Rokujo. Not only does this add a 'phantom' yet perfectly plausible addition to Genji scholarship in the tradition of Lovecraft's invented grimoire the *Necronomicon*, but it also proposes a subversive minor reading of a canonical work that foregrounds female agency. However, the 'weirdest' tendency in Enchi's novel is the way it 'veers' between different perspectives and points of mediation, transforming the reader's 'horizons of expectation' as it goes and frustrating any attempt at constructing an authoritative account of the past. While much of the narrative is focalized through one of Yasuko's suitors, the reader is also given long passages of Mieko's own writing and the lingering sense that the suitor is not really in control of his own inner world but is rather under the spell of those around him. What this alternation of perspectives leaves the reader with is a powerful sense of the relational construction of the characters involved, their inextricable interrelatedness. A section of Mieko's essay that describes the Lady Rokujo, and which is quoted by Ibuki in novel, uses spirit possession as a metaphor for relationality: 'Her spirit alternated constantly between spells of lyricism and spirit possession, making no philosophical distinction between the self alone and in relation to others, and unable to achieve the solace of a religious indifference.'[68]

The twins in *Masks* are the embodiment of the impossibility of distinguishing between the 'self alone' and the self 'in relation to others', one of the main effects of Enchi's veering narrative. But there is one further way in which the treatment of twins in the novel contributes to its weirdness. Despite attempts to use Harume as a form of haunted media, providing a line of communication with her dead twin, the channel that connects the siblings is ultimately depicted as blocked or inaccessible. If Mieko's scheme is an elaborate act of mourning, then that process is presented as a failure, as something that will never reach a satisfying point of closure. The finite practice of mourning is replaced by the ongoing, endless state of melancholy. In his 1917 essay 'Mourning and Melancholia', Freud characterizes melancholy as failed mourning. If healthy mourning is the process of the mourning subject severing (decathecting) the libidinal ties that bind the ego to the lost object (a dead loved one, for example), then melancholy results when this process fails and the lost object becomes lodged within the ego. If the temporality of mourning is progress (the ego severs from the past in order to continue into the future) then that of melancholy is perpetual recurrence, the strange loop. Haunting is replaced by weirdness. And it is the surviving twin Harume who embodies that state of melancholy in her retreat from the world. When Ibuki first catches sight of Harume at a summer party hosted by Mieko, he thinks she is from a different era and that time has become disjointed: 'Yes, that was just how she seemed: like a typical young lady of Meiji who had drifted into our times without aging a day. There was something unusual about her.'[69] Yasuko also thinks of Harume in terms of disjointedness or disconnection: 'There was something vaguely disturbing about her face, a sort of incoherence, as though the pitiable slumber of her mind had disconnected each vivid feature from the other.'[70] Harume, and the twin bond for which she provides access, are the embodiment of the blocked, 'disconnected' state of melancholy. When Ibuki is drawn into the melancholy of the Togano family, he grows 'fearful of losing his sense of time completely'.[71]

The documentary *Tell Me Who I Am*, directed by Ed Perkins, makes a similar connection between twins as a form of weird media and the state of melancholy. The film tells the story of 54-year-old twin brothers Alex and Marcus Lewis who are struggling to come to terms with the sexual abuse they suffered at the hands of their mother when they were children. The narrative constructed in the film closely follows that of the 2013 book, which was co-written by the brothers together with Joanna Hodgkin, and focuses on the relationship between the boys following a motorcycle accident that left Alex, at the age of eighteen, with acute amnesia. Unable to recall anything about his family or their past, but recognizing his twin brother, Alex reconstructs his life with the help of Marcus. The first 'Act' of the documentary tells the story from Alex's perspective, recounting how Marcus helped him piece his identity together, and ends with a disturbing revelation made while clearing out their mother's house after her death. Alongside other strange discoveries, including a pile of birthday presents addressed to the boys from family friends that they were never given, Alex finds a photograph in a locked draw of the two brothers when children, naked and with the heads torn off. When confronted with the photograph, Marcus finally admits that they had been abused and that he, as his brother's only trusted point of access to the past, had protected Alex from the painful elements of their family history and constructed for him a rose-tinted version of their childhood. Marcus tells his version of the story in 'Act 2' while the final section focuses on a dialogue between the two brothers.

The film evokes several gothic horror tropes. The house in which the Alex and Marcus grew up, Duke Cottage, is presented as being haunted by family ghosts. As the book puts it, '[g]radually, Alex discovered that his family was as full of shadowy corners as the old, rambling house they had grown up in'.[72] The house is the embodiment of the family psyche, full of generations of clutter and hiding places that the boys must work through as a corollary to their therapeutic narration. Furthermore, the twins are described in terms of gothic doubles who, like Dr Jekyll and Mr Hyde, represent two halves of the same personality. The book repeatedly emphasizes their identicalness describing them as 'always together, mirroring each other's gestures, laugh and voice'.[73] The film echoes this with the repeated use of a photograph, presumably taken not long before the accident, showing Alex and Marcus with the same haircuts and matching white shirts, as if mirrored reflections of one another. The uncanny is also an explicit theme in the book's treatment of the supposed 'psychic' connection that exists between the siblings. The reader is told that, before the writing of the book, the brothers featured in a daytime TV show that 'focussed on Alex's accident and the way Marcus had "known" about it at the same time: his premonition, and the uncanny way in which some twins seem able to communicate'.[74] Although it is not the focus of the film, the book dedicates considerable space to Alex's use of spiritualism to come to terms with the fall-out from his accident. A psychic tells him that the 'out-of-body' experiences he has been having are a form of 'astro-travelling'.[75] It is as if their twinship has predisposed them to an ability to slip in and out of bodies and inhabit other people's lives.

Much of this aligns Alex and Marcus's relationship with what Thacker terms 'haunted media'. Marcus mediates the past for his brother Alex. However, a closer inspection reveals the channel of communication to the past to be anything but clear. The film places a particular emphasis on the mediation of memory. After the accident, Marcus takes advantage of the potential of photography to construct false memories. Marcus would show Alex a photograph of the two of them on a beach and Alex would 'join the dots': 'He would give me a photo and I would construct a memory around that of two happy boys on a beach'. The 'tiny little fragments' provided by these photographs, inserted into a new context by Alex, 'became the building blocks for [Alex's] new sense of self'. The central role of these images in his new identity makes Alex obsessed with photographing his life after the accident due to his 'total paranoia that he would lose his new memory'. Alex was living inside images and they provided the parameters of his new existence. The focus on the technological mediation of memory in the film encourages the reader to scrutinize how the twins mediate each other's memories and experiences. Rather than communication with the past, this mediation is characterized by blockage. And the blockage is not just the one-way direction of Marcus blocking Alex's access to the past. In Marcus's account, Alex also becomes a repository for the false memories that he has implanted there to the point that the artificiality of these memories is forgotten. As the book puts it, '[f]or Marcus, the dividing line between fact and fiction had often been blurry. He'd been acting as the custodian of Alex's memories, but most of the time his grasp on his own felt uncertain'.[76] In a twist on the cybernetic twins trope discussed in Chapter 1, the mediation between the brothers is characterized by a feedback loop. Marcus implants the false memories in Alex and Alex feeds them back to Marcus, stripped of their falseness and bearing every resemblance to truth. As Marcus explains in the film, 'I began to believe the reality that I created. [...] Alex lost his memory by accident. And I lost my memory voluntarily'. The brothers present themselves as melancholy twins who are locked in perpetual mourning for the versions of themselves they lost to child abuse.

While both the book and the film present themselves as a form of therapy for the boys, a vehicle through which to process their trauma, they identify different objectives for this procedure. The book is primarily concerned with enabling Alex and Marcus to finally 'separate' from each other. 'If their lives had followed a more normal path, they would have begun to separate in their teens. [...] But his accident, and his total dependence on his twin, interrupted their efforts to separate.'[77] The twins' failure to separate, their lack of individuality, is presented as a pathological symptom that the book, and the therapeutic process of confession behind the book, sets out to cure. This is emphasized by the fact that, although the majority of the book has been written by Joanna Hodgkin based on the Alex and Marcus's words, it is bookended by a prologue and epilogue authored by the brothers separately. The implication is that, thanks to the book, they have finally found their individual voices. The film presents it differently. Although the first two 'Acts' focus on the separate perspectives of the two brothers, the third places them in dialogue. And while it does perform a process of closure (albeit not very convincingly) the subjectivity it leaves the viewer with is an entwined one. The brothers are more together than ever. As Alex puts it, 'It's over, finally. And *we* move on' [My emphasis]. Rather than purely mutually destructive, their shared and mutually enforcing melancholy has been productive of an entwined subjectivity similar to the 'self in relation to others' produced by the image of twinship constructed in *Masks*.

Weirding genetic transmission

There are clear parallels between the narrative device of twins maintaining communication beyond death (twinship as haunted medium) and the use of twinship in the field of behavioural genetics. In both, twinship is constructed as a medium that is characterized by 'transparency'. In behavioural genetics research, twins are thought to provide a window onto the determining forces of genetic material. Reviews of psychologist and geneticist Robert Plomin's 2018 book *Blueprint: How DNA Makes Us Who We Are* were accompanied by photographs of identical twins, exposing the frequency with which images of twinship are used as a visual shorthand for genetic determinism.[78] In several recent horror films, twinship has also been used to explore the power dynamics surrounding the conception of the human body as a haunted medium. Films such as David Cronenberg's *Dead Ringers* (1988) and Peter Greenaway's *A Zed & Two Noughts* (1985) remind the viewer of the biopolitical contexts of using twins as objects of study. As Elana Gomel and Stephen Weninger point out, both films make the historical connection between biomedical research and twinship in different ways. In both films, twinship is the object of scientific intervention and evokes the spectre of eugenics.[79] In *Dead Ringers*, this is carried out through the alignment of fertility treatment and patriarchal power structures. By placing twins as both spectacles and objects of scientific interest within a zoo, *A Zed & Two Noughts* (discussed at greater length later in the chapter) evokes the symbiotic relationship between capital and biopolitical interests. In this context, 'weirding' the medium of twinship by rendering it opaque and denying its transparency both disrupts the conscription of twins into scientific research and points to the ideological continuities between the eugenics movements of the first half of the twentieth century and contemporary genetics research.

Before returning to Greenaway's work, it is worth looking at how the weirding of twinship as a medium is carried out in a film that positions itself more squarely within the horror genre: *The Unborn*, directed by David Goyer and released in 2009. The film tells the story of Casey, a high-school girl who is haunted by the spirit of her grandmother's twin brother who died while undergoing experiments at the hands of Josef Mengele in Auschwitz. Following a series of unsettling experiences, including dreams about a live baby bottled in formaldehyde and being attacked by a six-year-old with the sharp edge of a broken mirror, Casey notices that her eyes are starting to change colour. After running a few tests, the ophthalmologist suspects that Casey's *heterochromia iridis* has been caused by genetic mosaicism, the transfer of genetic material while in utero. 'Are you a twin by any chance?' he asks. Research into the family archives reveals that Casey did in fact have a twin who died in childbirth and that the nickname her parents used for him was also used by the six-year-old before he attacked her: 'Jumby wants to be born now.' Her grandmother tells her that in Auschwitz an evil spirit called a dybbuk used her twin brother Barto's body as a gateway into the physical world. Now, it seems, the spirit has taken control of Casey's unborn twin brother and is trying to stage a return.

Twinship in the film is constructed as a medium. When Casey learns about Barto she observes: 'some people are doorways'. 'Yes', replies her grandmother, 'especially twins'. The treatment of twinship as a medium hesitates between the two categories outlined by Thacker. On the one hand, twins are presented as a haunted medium. The haunting carried out by the dead twins Barto and Jumbi literalizes the use of twins in eugenics. For Nazi doctors, by exposing the influence of genetic inheritance on human development, twins render visible the agential role of the racial past in the formation of identity in the present. By haunting the present, the dybbuk is the nightmare underside of the Nazi dreams of racial purity. As well as twin bodies, the dybbuk manifests itself through digital media devices. The first moment of horror in the film takes place while Casey is babysitting for her neighbours and hears the six-year-old, inhabited by the dybbuk, through the static of the baby monitor. From then on, the dybbuk is mostly glimpsed refracted through various forms of visual media, whether it be the video-call window on Casey's laptop monitor or the images cast by the super 8 footage found in the dusty family attic. But the medium of twinship is also disrupted and rendered weird in the film. The proliferation of forms the dybbuk takes in *The Unborn* and repeated attempts to account for its existence never get the viewer any closer to identifying what it actually is. The opacity of the menacing force at the centre of the film provokes the types of 'weirdness' identified by Harman in the work of Lovecraft. The representation of the dybbuk is weird in a 'horizontal' sense in that not even the experts consulted by Casey claim to know what it is. Both the Rabbi and the Christian exorcist who help her during the schlocky denouement claim no knowledge of the nature of the evil entity but rather concern themselves with the effects it produces in the material world. The representation of the dybbuk is also weird in the 'vertical' sense since it seems to compensate for this fundamental lack of knowledge (this gap) with an excessive proliferation of descriptions. The film is assembled from a disjointed montage of different references to genre conventions (there are nods to a range of iconic horror films, from *The Shining* to *The Exorcist*) and Jewish mysticism (including to a Kabbalistic *Book of Mirrors*).

The affective tenor of the film hesitates between the uncanny and the weird. The two modes clash in an uncomfortable way in a sequence that takes place in the bathroom stalls of a nightclub. The narrative content of the sequence is uncanny. Casey is visited by the ghost of her dead mother whose fate she seems to be repeating. But the aesthetics of the sequence are straight out of weird fiction. The sides of Casey's stall are penetrated by monstrous tentacles (that staple Lovecraftian monstrous appendage; described by Luckhurst as the 'emblem of that which will not correlate, be reduced to human thought') the existence of which is never given a narrative explanation.[80] Uncanny and weird mediation collide in the last sequence of the film. In a final twist, we learn that, like her mother and her grandmother before her, Casey is pregnant with those monstrous portals of evil: twins. As the stenographer announces that 'it seems that double congratulations are in order' we cut to the six-year-old repeating the ominous line 'Jumby wants to be born now.' Just before the credits start to role, the camera zooms into the sonogram monitor showing the twins in utero as the sounds of their twin heart beats crescendo to block out all other noises. The emphasis on the audio track in this sequence is significant since, unlike the other visual media employed within the film, the main active agent in sonographic imaging is not light but sound. In sonography, the foetus is made visible through the vibration of ultrasound waves through the womb and the conversion of the information produced by these vibrations into a digital image. As the camera zooms into the sonogram monitor, the muffled, sub-aquatic throbbing sounds seem to visibly cling to the photographs, the noises transmuted into the pixelated blur of tissue mass as in-utero space looms into and out of visibility. As Kristopher Cannon puts it in his account of the sonographic image, 'a new picture flickers with every pulse of ultrasound, images transform into other images [...] every visible trace becomes a trace of un-visibility.'[81] Sonogram images 'flicker within thresholds between static forms of visibility [...] making this an intrinsically unstable space'.[82] Like twinship itself sonographic imaging operates on the boundary between visibility and opacity.

This reminder of the resistance of uterine space to technologies of visuality draws attention to how the film carries out a weirding of twin mediation at the level of plot. Key here is Casey's diagnosis by the ophthalmologist. Perhaps because he is an ophthalmologist and not a genetic counsellor, his diagnosis is slightly confusing. While he uses the term 'genetic mosaicism', which is when two or more populations of cells with different genotypes are found in one individual developed from a single zygote, the process he is actually describing is 'chimerism', when two genotypes arise from the fusion of more than one zygote during embryonic development. Chimerism is one of a number of processes that take place during early embryonic development cited by critics of twin methods. The exchange of DNA between embryos in utero would seem to undermine the distinction between monozygotic and dizygotic twinship that constitutes the theoretical keystone of twin studies research. Chimerism is cited by geneticist Charles Boklage, often described as a 'maverick' in the field, as evidence for why some of the fundamental premises of twin biology are wrong.[83] Whether or not you believe some of Boklage's more speculative conclusions, the occurrence of chimerism renders twinship a much less transparent medium of genetic inheritance than is often acknowledged by behavioural geneticists. The message that *The Unborn* leaves us with is that twinship is a much weirder form of mediation than its use in the popularization of genetics research would leave us to believe.

Conclusion: (In)organic mediation

The narratives studied in this chapter present twinship as a form of mediation that blurs the distinctions between the organic and the inorganic, the animate and the inanimate. This is a quality shared by all the texts examined here that incorporate twins into narratives that operate at the fringes of the horror genre. It is also true of the use of twins in genetics research. William Viney argues that, when conscripted into scientific research, twins are made to operate across scales running from animate to inanimate. 'Their recruitment into scientific inquiries is brought about and validated by being both sentient human beings and inert, malleable, disposable things.'[84] As tools of the MISTRA project, the Springer twins, discussed in the Introduction, were both exploited as a headline-grabbing human drama story and reduced to a subhuman level of genetic information. Films such as *The Unborn* make this cross-scale use of twins explicit by positioning twinship as a medium that connects distinct ontological domains. By way of conclusion, I want to turn to an earlier film, *A Zed & Two Noughts* (1985), directed by Peter Greenaway, to examine how it presents twinship as a form of mediation that blurs the distinctions between the organic and the inorganic that undergirds anthropocentric enlightenment thinking. Greenaway's film reveals a key tendency of this shift from a model of uncanny to weird twinship that I have been tracing throughout this chapter. Whereas uncanny twins, following Poe's paradigm, are a product of introspective entropic decay and death, weird twinning is a process of monstrous and unpredictable nonhuman becoming.

The narrative of the film focuses on zoologists Oswald and Oliver Deuce, formerly conjoined twins who were surgically separated as babies. Grief-stricken by the deaths of their wives, they become obsessed with the decay of organic matter and, in an attempt to come to terms with their loss (and loss in general), conduct a series of experiments using photographic technologies to scrutinize its processes. In another apparent emotional reaction to grief, the brothers gradually attempt to return to their conjoined state, first getting the same haircuts and dressing the same and finally binding themselves together in double suits and shirts that evoke those worn by the 'original Siamese Twins' Chang and Eng. In their final experiment, the twins render themselves the objects of scientific scrutiny by injecting one another with poison and rigging up a camera to record their death and physical decay. [See Figure 2.1] The focus of *A Zed & Two Noughts* is the role of twinship in biopolitical power and the scopic regimes that reinforce it. Most critics of the film have interpreted the

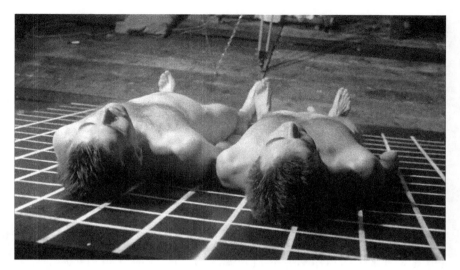

Figure 2.1 Oswald and Oliver Deuce make themselves the objects of their own experiment in *A Zed & Two Noughts* directed by Peter Greenaway, BFI 1985. All rights reserved.

suicide of the Deuce twins as an iteration of the Narcissus myth. In Ovid's version of the myth, Narcissus, captivated by the allure of his own image as it is reflected back to him in a pool of water, is consumed by introjected passion and dies. The replacement of an external love object with self-love leads to stasis and death. In *A Zed & Two Noughts*, the Deuce twins could be seen to stand in for both Narcissus and his reflection. By severing the bonds from their external love objects, the death of their wives has led them to, in Freud's terms, recathect their libidinal attachments onto each other. According to this reading, then, Oswald and Oliver become an iteration of Poe's gothic twins whose inward-looking connectedness leads them on a spiral of entropic decay.

Elana Gomel and Stephen Weninger argue that the 'return' to a 'static le-même-au-même symmetry of "ideal" twinship' by the Deuce brothers functions as symbolic compensation for their failure to scientifically understand the decay of the body.[85] 'Finding no plan or purpose in the body's senescence and unable to escape the torment of the asymmetrical, the twins seek to recapture an original wholeness in the perfect equipoise of death.'[86] The brothers' desire to capture their own deaths on film is an attempt to 'arrest the inexorable process of decay'.[87] Gomel and Weninger point out that the film places this ideal of twinship within the historical context of biopolitics. This context is evoked metaphorically through the three main spaces constructed in the film: the zoo, museum and laboratory. Like the other animals in the zoo, the Deuce brothers become a spectacle of consumer entertainment. The history of twinship as a spectacle is evoked both through the Chang and Eng suit that they wear in the final sequence and through a key scene in which both brothers enter a cage in front of a gaggle of gawking journalists. As a form of living museum, the zoo becomes a space in which the twins are categorized and controlled. The connection between zoo and museum is emphasized in a highly stylized manner typical of Greenaway, through repeated references to Vermeer. The final sequence, in which the twins become both subject and object of a scientific gaze, points to the historical role played by twins in constructing the body as knowable and controllable. Although *A Zed & Two Noughts* does not explicitly make reference to fascism, through these references it points to the historical entanglement between twinship and what Gomel and Weninger describe as 'the deadly corporeal ideologies of purity and sameness'.[88] The reinforcement of sameness by Oswald and Oliver evokes the appeal of twins to Nazi eugenics. By putatively proving the power of heredity, the twinned body 'provides the reassurance of self-identity' and in the process 'supplant[s] racial, sexual and individual difference with the economy of the Same'.[89] As both subject and object of scientific scrutiny ('both Mengele and his victims') the characters of the Deuce brothers emphasize how the 'negative eugenics' of murder and the 'positive eugenics' of cloning or twinning are two sides of the same coin.

The Deuce twins, according to this reading, are a variation on the melancholic twins motif we found in *Masks* and *Tell Me Who I Am*. And it is through the use of a discourse and aesthetic of melancholy that *A Zed & Two Noughts* 'weirds' the medium of twinship and diverts it from its historical role as an instrument of biopolitical power. Oswald and Oliver's desire to return to a state of corporeal conjoinment is presented as a physical corollary of their shared melancholy, which expresses itself as a mutual inward-looking 'stuckness'. In their grief, the brothers turn to each other and become one another's partially internalized substitute love objects. Furthermore, the melancholia of the two protagonists seems to infect the medium of film itself through its aesthetic of blocked mediation. Through its frequent use of symmetrical *mise en scène* and its detached long-distance framing, *A Zed & Two Noughts* implicates itself in the scientific scopic regime that it is critiquing at the level of the narrative. If, at the level of the plot, photographic technologies of the scientific gaze fail to impose order on the processes of decay, then the film also presents itself as a failure. Oswald and Oliver's attempt to film their own death and decay and, in the process, immortalize it as art is undone when snails overrun the camera and short its electronic circuits. The melancholic implosion of the twins (their fatal mutual internalization) is accompanied by an implosion of the medium itself. The weird mediation of twinness is conjoined with a weirding of the cinematic medium. At the end of *A Zed & Two Noughts*, Greenaway leaves the viewer with a model of cinema as failed or blocked mediation.[90]

However, alongside this melancholic blockage, *A Zed & Two Noughts* emphasizes a productive dimension to the weird mediation of twins. It is this productivity that critics such as Gomel and Weninger miss in their focus on the film's staging of an atemporal 'geminate ideal' of twinship. The affirmative nature of weird twinship in the film results from its alignment of melancholy with a Deleuzian rather than a Freudian interpretation of the death drive. In his essay 'The Ego and the Id', published in 1923, Freud observes that in melancholia we find 'a pure culture of the death instinct'.[91] Here, he makes a connection between his characterization of melancholy as failed mourning in his essay of 1917 and his account of the death drive in his 1920 work *Beyond the Pleasure Principle*. The experience of living, as Freud characterizes it in this work, is a compromise between the desire to sustain life and the urge for death. Life, as he sees it, is fundamentally entropic in its desire to return to the lowest possible energy state. Or, as the final phrase of the essay puts it, '[t]he goal of all life is death'.[92] For Freud, the life of the organism is closed off from the world, running down its own energy. In its closed-off introspection and its gradual tendency towards inertia, it is easy to see how Freud saw melancholy as the culture of the death instinct in its 'purest' form. At one level, *A Zed & Two Noughts* echoes Freud's connection between melancholy and the death drive. The decaying animal carcasses filmed by Oliver Deuce function as a metaphor for the twins' grief-induced melancholic psychic economy.

However, what emerges from the association between melancholy and the death drive – established through the film's construction of weird twinship – is not a Freudian entropic vision of life as a tendency to death but rather a conception of life as an affirmative struggle against decay. The film is interspersed with the time-lapse images of decaying animals (a fish, crocodile and zebra) taken by Oliver. While these shots make explicit the complicity of cinema itself in the scopic regime that undergirds eugenics, they also draw attention to the liveliness of organic decay. The images of decomposing animals present the carcasses as sites of resurgent life as they become animated by the activity of maggots, first seething under the skin and then oozing from the eyes. Rather than stasis, death is presented in these time-lapse images as a transition from one state to another. In this, it echoes Deleuze's characterization of the death drive in his discussion of *Beyond the Pleasure Principle* in *Difference and Repetition*, as the collapse of a given structure in the face of pure becoming. As Keith Ansell-Pearson puts it in his exploration of the biophilosophy that informs Deleuze's wider project to 'think "beyond" the human condition', dissolution and decomposition are configured 'in terms of the vital and virtual role they play within the creative evolution of matter and of complex systems'.[93] Freud construed the repetition constitutive of the death drive in terms of a regression or 'return to inanimate matter' due to his 'commitment to a conception of the personal unity and integrity of the organism'.[94] In Deleuze's re-working of the concept of the death drive, repetition holds a 'demonic' power, existing as it does between the borders of the living and the dead: 'If repetition makes us ill, it also heals us; if it enchains and destroys us, it also frees us.'[95] In Ansell-Pearson's words, for Deleuze, 'repetition, including the death-drive, is involved in an emancipation of the subject from its own repressive character'.[96] Although in many ways evocative of the Freudian understanding of death drive (the return to inanimate state of death is accompanied by a regression to the childhood state of conjoinment), *A Zed & Two Noughts* ultimately affirms this latter unsettling reframing of the concept. The repetition of twinship in the film is not the comforting repetition of heredity. Rather, twinship in the film evokes the 'demonic' power of the transversal movements of material forces and effects that which Deleuze terms the plane of immanence. As with all the texts examined in this chapter, Greenaway's films oppose the translucent mediation of twinship as it is used in genetics research with a model of twinship as a weird mediation that blocks the scopic regimes of modernity while conjoining realms kept separate by this regime: life and death, organic and inorganic, animate and inanimate.

Notes

1   Sigmund Freud, 'The Uncanny', in *The Essentials of Psychoanalysis*, ed. Anna Freud and trans. James Strachey (London: Pelican, 1986 [1919]), 366.

2   For an account of the association between electronic media and paranormal phenomena see Jeffrey Sconce, *Haunted Media: Electronic Presence from Telegraphy to Television* (Durham, NC: Duke University Press, 2000).

3   Eugene Thacker, 'Dark Media', in *Excommunication: Three Inquiries in Media and Mediation*, ed. Alexander R. Gallway, Eugene Thacker and McKenzie Wark (Chicago and London: The University of Chicago Press, 2014), 81.

4   Thacker, 'Dark Media', 89.

5   Thacker, 'Dark Media', 95.
6   Thacker, 'Dark Media', 95.
7   Edgar Allan Poe, *Tales of Mystery and Imagination*, ed. Graham Clarke (London: Everyman Library, 1984 [1839]), 141.
8   Poe, *Tales of Mystery and Imagination*, 149.
9   Poe, *Tales of Mystery and Imagination*, 6.
10 Poe, *Tales of Mystery and Imagination*, 142.
11 Poe, *Tales of Mystery and Imagination*, 138.
12 Bram Stoker, *The Dualitists; or the Death Doom of the Double Born* (Auckland: The Floating Press, 2016 [1886]), 8.
13 Stoker, *The Dualitists*, 21.
14 Stoker, *The Dualitists*, 11.
15 Stoker, *The Dualitists*, 21.
16 Stoker, *The Dualitists*, 24.
17 The 2011 horror film *Seconds Apart*, from low-budget indie horror producers After Dark Films, explores the association between twins, entropy and cybernetics. The film tells the story of Seth and Jonah Trimble, misanthropic teen twins with psionic abilities that include a powerful telepathic bond and telekinetic powers that can radically intervene into other people's experience of reality. Their intense connection with one another seems to have come at the price of wider social inclusion. They are shunned by their classmates as 'freaks' and return this antipathy with psychotic violence. The opening scene takes place in a typical high-school house party in which a group of jocks play a drinking game in an upstairs room. Events take a sinister turn when the beer bottles are replaced with a handgun and, without the fear you might expect, the game turns into a round of Russian roulette. The viewer subsequently discovers that the whole scene had been filmed and orchestrated by Seth and Jonah using their mind control abilities. The second sequence shows us the brothers, played by monozygotic twins Edmund and Gary Entin, watching their recording of the staged suicide on a computer screen. They play back the moment of death repeatedly and reluctantly conclude that they 'don't feel anything'. The murder, we learn, was part of their 'project', the aim of which is never explicitly stated but seems to relate to empathy. The boys commit acts of horrendous violence on those around them, record them on video and watch them back in an attempt to test the limits of their social detachment and sound out the possibility of empathic connection with those around them. Significantly, the attempts at empathic communication are always mediated by communications technologies. They do not question whether they feel something 'in the moment' but rather when watching the footage back on a screen. As a result, the 'project' seems to be testing not just their own capacity for empathy but that of the media assemblage of which they form a part. The construction of

the twins as media cyborgs is emphasized throughout the film by their ability to manipulate communications technologies. One of their acts of violence takes place in a language lab in which Seth and Jonah's fellow pupils listen to Spanish phrases and repeat them back into a microphone. In an attempt to deter a classmate from reporting them to the police, the boys take control of the headset to deliver their message. Their media manipulation is just one example of how the twins induce hallucinations in those around them to control the border between visibility and invisibility. At the beginning of the film Seth and Jonah are presented as a closed system. Their closeness with each other (born only 'seconds apart') has produced around them an almost impenetrable bubble. The narrative charts the disintegration of this bubble and their transition from a closed system to an open system.

18 Nicholas Royle, *The Uncanny* (Manchester: Manchester University Press, 2003), 2.
19 Christopher Johnson, 'Ambient Technologies, Uncanny Signs', *Oxford Literary Review* 21 (1999): 128.
20 Johnson, 'Ambient Technologies, Uncanny Signs', 125.
21 David Lynch, *Lost Highway* (USA: Asymmetrical Productions, 1997).
22 H.P. Lovecraft, 'The Dunwich Horror', in *The New Annotated H.P. Lovecraft*, ed. Leslie S. Klinger (New York and London: Liveright Publishing Corporation, 2012 [1929]), 348.
23 Lovecraft, 'The Dunwich Horror', 346.
24 Lovecraft, 'The Dunwich Horror', 352.
25 Lovecraft, 'The Dunwich Horror', 356.
26 Lovecraft, 'The Dunwich Horror', 358.
27 Graham Harman, *Weird Realism: Lovecraft and Philosophy* (Winchester and Washington: Zero Books, 2012), 5.
28 Harman, *Weird Realism*, 4.
29 Harman, *Weird Realism*, 24.
30 Harman, *Weird Realism*, 25.
31 Roger Luckhurst, 'The Weird: A Dis/Orientation', *Textual Practice* 31:6 (2017): 1050.
32 Luckhurst, 'The Weird', 1052.
33 Luckhurst, 'The Weird', 1052.
34 Lovecraft, 'The Dunwich Horror', 386.
35 Otto Rank, *The Double: A Psychoanalytic Study*, trans. Harry Tucker, Jr. (Chapel Hill: University of North Carolina Press, 1971), 3.
36 Mark Fisher, *The Weird and the Eerie* (London: Repeater Books, 2016), 10–11.
37 Jeff Vandermeer, *Acceptance* (London: Fourth Estate, 2014), 8.
38 Vandermeer, *Acceptance*, 15 and 292.
39 Vandermeer, *Acceptance*, 30–1.

40 Vandermeer, *Acceptance*, 31.
41 Vandermeer, *Acceptance*, 37.
42 Sara Wasson, 'Love in the Time of Cloning: Science Fictions of Transgressive Kinship', *Extrapolation* 45:2 (2004): 140.
43 Vandermeer, *Acceptance*, 193–4.
44 Vandermeer, *Acceptance*, 194–5. Italics in original.
45 Vandermeer, *Acceptance*, 194.
46 Vandermeer, *Acceptance*, 195.
47 Vandermeer, *Acceptance*, 193.
48 Lovecraft, 'The Dunwich Horror', 344.
49 Patricia Welch, 'Excess, Alienation and Ambivalence: Edogawa Rampo's Tales of Mystery and Imagination', in *Japanese Tales of Mystery and Imagination*, trans. James B. Harris (Tokyo: Tuttle Publishing, 2012), 14.
50 Shimezu Yoshinori quoted in Baryon Tensor Posadas, 'Rampo's Repetitions: The Doppelganger in Edogawa Rampo and Tsukamoto Shin-ya', *Japan Forum* 21:2 (2009): 161.
51 Edogawa Rampo, 'The Twins', in *Japanese Tales of Mystery and Imagination*, trans. James B. Harris (Tokyo: Tuttle Publishing, 2012 [1956]), 144.
52 Rampo, 'The Twins', 150.
53 Allan Sekula, 'The Body and the Archive', *October* 39 (1986): 3–64.
54 Lisa Zunshine, *Strange Concepts and the Stories They Make Possible* (Baltimore: The Johns Hopkins University Press, 2008), 35.
55 Zunshine, *Strange Concepts*, 35.
56 Thacker, 'Dark Media', 129.
57 Thacker, 'Dark Media', 133.
58 Thacker, 'Dark Media', 134.
59 Niffenegger's bestselling first novel *The Time Traveller's Wife* (2003) pulls off a similar trick by inserting a hackneyed science fiction convention within a psychologically realist romance plot.
60 Fumiko Enchi, *Masks*, tran. Juliet Winters Carpenter (New York: Vintage Books, 1983 [1958]), 10.
61 Enchi, *Masks*, 24.
62 Enchi, *Masks*, 24.
63 Doris G. Bargen, 'Twin Blossoms on a Single Branch: The Cycle of Retribution in *Onnamen*', *Monumenta Nipponica* 46:2 (1991): 151–2.
64 Enchi, *Masks*, 13.
65 Enchi, *Masks*, 70–1.
66 Enchi, *Masks*, 72.
67 Luckhurst, 'The Weird', 1048–9.
68 Enchi, *Masks*, 132.

69 Enchi, *Masks*, 41.
70 Enchi, *Masks*, 70.
71 Enchi, *Masks*, 84.
72 Alex and Marcus Lewis with Joanna Hodgkin, *Tell Me Who I Am* (London: Hodder & Stoughton, 2013), 29.
73 Lewis with Joanna Hodgkin, *Tell Me Who I Am*, 11.
74 Lewis with Joanna Hodgkin, *Tell Me Who I Am*, 112.
75 Lewis with Joanna Hodgkin, *Tell Me Who I Am*, 120.
76 Lewis with Joanna Hodgkin, *Tell Me Who I Am*, 325.
77 Lewis with Joanna Hodgkin, *Tell Me Who I Am*, 235.
78 Robert Plomin, *Blueprint: How DNA Makes Us Who We Are* (London: Allen Lane, 2018).
79 Elana Gomel and Stephen Weninger, 'Cronenberg, Greenaway and the Ideologies of Twinship', *Body & Society* 9:3 (2003): 19–35.
80 Luckhurst, 'The Weird', 1054.
81 Kristopher L. Cannon, 'Ec-statically Queer Images: Queering the Photographic through Fetal Photography', *Photography and Culture* 7:3 (2014): 278.
82 Cannon, 'Ec-statically Queer Images', 278.
83 Charles E. Boklage, *How New Humans Are Made: Cells and Embryos, Twins and Chimeras, Left and Right, Mind/Self/Soul, Sex, and Schizophrenia* (Singapore: World Scientific Publishing, 2010).
84 William Viney, 'Getting the Measure of Twins', in *The Edinburgh Companion to the Critical Medical Humanities*, ed. Anne Whitehead, Angela Woods, Sarah Atkinson, Jane Macnaughton and Jennifer Richards (Edinburgh: Edinburgh University Press, 2016), 110.
85 Gomel and Weninger, 'Cronenberg, Greenaway and the Ideologies of Twinship', 25.
86 Gomel and Weninger, 'Cronenberg, Greenaway and the Ideologies of Twinship', 27.
87 Gomel and Weninger, 'Cronenberg, Greenaway and the Ideologies of Twinship', 27.
88 Gomel and Weninger, 'Cronenberg, Greenaway and the Ideologies of Twinship', 30.
89 Gomel and Weninger, 'Cronenberg, Greenaway and the Ideologies of Twinship', 28 and 29.
90 Brian de Palma's 1972 psychological slasher film *Sisters* employs a similar strategy. The film centres on formerly conjoined twin Danielle who was separated from her sister Dominique in a high-profile experimental surgery. In a news interview, the head surgeon of the institute carrying out the procedure warns that the pair were brought up as one and that separation might 'disturb their psychological balance'. This prediction is proved correct when the more 'disturbed' of the two, Dominique, dies not long after separation and the affable Danielle internalizes her sister's disturbances. The narrative focuses on Danielle failing to take her medication and her monstrous other half 'Dominique' murdering a man after a one-night stand.

The appearance of the doubling motif is accompanied by an attack on cinematic conventions, starting with the use of split screen following the murder and culminating in a delirious and disorienting dream sequence.

91 Sigmund Freud, 'The Ego and the Id', in *The Standard Edition of the Complete Psychological Works of Sigmund Freud (Vol. XIX)*, ed. James Strachey (London: Hogarth Press, 1923), 53.

92 Sigmund Freud, 'Beyond The Pleasure Principle (1920),' in *The Standard Edition of the Complete Psychological Works of Sigmund Freud (Vol. XIV)*, ed. James Strachey (London: Hogarth Press, 2001 [1957]), 7–66.

93 Keith Ansell-Pearson, *Germinal Life: The Difference and Repetition of Deleuze* (London and New York: Routledge, 1999), 81.

94 Ansell-Pearson, *Germinal Life*, 98 and 99.

95 Gilles Deleuze, *Difference and Repetition*, trans. Paul Patton (London: Bloomsbury, 2014), 21.

96 Ansell-Pearson, *Germinal Life*, 105.

3

Twins in the Anthropocene

Twins have long been used to conceptualize connections between human societies and the natural world. Early anthropological narratives teem with accounts of twins acting as mediators between humans and animals or humans and the elements. In *The Golden Bough*, James George Frazer explores 'a widespread belief' among the Indian tribes of British Colombia that 'twin children possess magical powers over nature, especially over rain and the weather'.[1] The Tsimshian Indians, for example, attempt to control strong winds with the exhortation: 'calm down, breath of the twins'.[2] For the Kwakiutl Indians, twins are anthropomorphized salmon and are prevented from nearing water for fear that they might transform back again and swim away. Because of this proximity with the natural world, they play an important mediating function, and are both feared and revered for their abilities to 'summon any wind by motions of their hands', 'make fair weather or foul weather' and 'cure diseases by swinging a large wooden rattle'.[3] Such beliefs, Fraser points out, are also widespread in other parts of the world. The Baronga tribe of Southeastern Africa, for instance, calls twins 'children of the sky' and ascribes to them the power to reverse drought.[4] Far from a peculiarity of Victorian anthropologists, this fascination with the connection between twins and nature has endured throughout the twentieth and twenty-first centuries to become a conceptual lens through which to view entanglements between humans and their environments in the context of the mounting climate crisis.

Within this context, twinship has been mobilized within two contrasting narratives. The first narrative presents twins as embodying technological control over nature. In *The Ends of the World*, Débora Danowski and Eduardo Viveiros de Castro trace the emergence of a number of mythological discourses that have developed in response to the spectre of human extinction. Dominant among these are: the image of a world *before* humans or before the modern separation between human and world; the projected vision of a world *after* humans as a 'restitution of the world to itself'; the idea of a future in which the world will have become 'human' due to large-scale environmental collapse; and, finally, the notion of a world that has been 'transubstantiated and absorbed by humankind as the triumphant species that re-transcends itself' through 'ingenious feats of anthropo-engineering'.[5] It is as part of this latter narrative that twins perform a starring role. Evoking the nightmare of Nazi eugenics, popular culture has repeatedly affirmed the connection between twins and clones. The self-replication that takes place in monozygotic twins has been taken as evidence of the predisposition of the natural world to control and manipulation. In this, twins-qua-clones embody the enduring fantasy of the 'technical fix', the possibility that if technological modernity has gotten us into this situation it can get us out again. The repetition embodied by twins stands in, not for the cancellation of the future through endless repetition of the present, but for the emergence of a 'worldless humanity' in which 'the anthropic dreams of the Moderns would [...] be finally materialized: a post-environmentalism in which man will find himself contextualized and sustained only by himself'.[6] A number of the mid-twentieth-century science fiction narratives discussed in Chapter 1 use the association between twins and eugenics to pre-empt this vision of a 'good Anthropocene' in which the Earth's natural rhythms are fully tamed and ordered. These stories echo those of Frazer in which twins are agents of the magical containment of environmental excesses. The promethean promises of geo-engineering are modernity's corollary of the 'calm down, breath of the twins' prayers of the Tsimshian people.

The second discourse into which twins have been conscripted is one of ecological entanglement rather than human exceptionalism. This is a variation of Danowski and Viveiros de Castro's 'before' and 'after' myths and positions twinship as a model for a non-modern engagement with the world. Echoing the role of twinship in the performance of cybernetic ontologies, this second discourse contends that the recursion of the developmental interactions between twins predisposes them to an ecological openness. The cybernetic performativity of twins explored by William Grey Walter and Norbert Wiener paved the way for what Uni Chaudhuri has described as 'anthropoScenic' performances.[7] This confluence in the use of twins points to the fact that, as Yuk Hui puts it, '[e]cology, beyond its strict use in biology, is not a concept of nature but rather a concept of cybernetics'.[8] In a description of transversal connections between art and science in eco-critical practice, Donna Haraway (2017) argues that, as twin sisters, activists Margaret and Christine Wertheim were 'primed for sympoietic mergers'.[9] It is a seemingly throwaway comment in an essay that draws parallels between 'symbiogenetic' processes at the level of cells and organisms and environmental activism that combines artistic and scientific approaches. However, the statement is indicative of a discourse running through a range of disciplines and representational practices that positions twins – particularly monozygotic or so-called 'identical' twins – as a model of trans-corporeal subjectivity commensurate with the conceptual reconfigurations required by the Anthropocene. The experience of being twins, Haraway suggests, has prepared the Wertheim sisters for the forms of sympoietic 'making with' necessary for a critical conceptualization of the entanglements of the human with its non-human others.

Narratives about twins have emerged with particular force in relation to the set of debates centring on the current climate crisis. Literary fiction frequently has recourse to twin characters to break down oppositions between nature and culture and, most frequently, between the human and the animal. Ágota Kristóf's *The Notebook* (1986) and Gerbrand Bakker's *The Twin* (2006) both construct extended parallels between the affective and bodily entanglements between twins and interrelations between the human characters and the natural world. In both cases, the demands of coming to terms with the entwined being of twins is presented as having predisposed them to an openness to experiencing subjectivity as something that is in a constant state of construction through recursive interactions with the environment. *The Notebook* (the first in a trilogy that continues with *The Proof* in 1988 and *The Third Lie* in 1991) recounts the exploits of young twin boys who live in a small Hungarian town during the final years of the Second World War and the first years of the communist regime. The first book of the trilogy is a tale of survival following the boys' abandonment by their mother and the presumed death of their father at the front. Sent to live with their peasant grandmother, the narrative charts a dual trajectory as the twins merge both with the land and with each other. The comparison between the boys and the natural world is established in the opening chapter as the cruel grandmother, on discovering that she has twin grandsons, mocks her daughter by asking where the others are: 'Bitches have four or five puppies at a time. You keep one or two and drown the others'.[10] The grandmother's taunt, which draws on a long-standing convention of characterizing multiple births as animalistic,[11] foreshadows the transformation of the boys as they lose the trappings of their civilized pre-war life in the city: 'We're getting dirtier and dirtier [...]. Where possible, we walk barefoot [...]. The soles of our feet are getting hard [...].'[12] In an attempt to prepare themselves for the coming hardships, the boys begin to carry out a series of 'exercises to toughen the body' that include hitting and burning each other.

As well as a process of merging with dirt, the transformation undergone by the twins is described in terms of mutual entwinement. As part of their training, one of the twins pretends to be blind while the other pretends to be deaf. After a while, they no longer require 'a shawl over our eyes or grass in our ears' since 'the one playing the blind man simply turns his gaze inwards and the deaf one shuts his ears off to all sounds'.[13] They become one body in an apparent confirmation of their mother's insistence that they 'are one and the same person'.[14] This radical embrace of twinship and refusal of separation is described as a form of desubjectification and psychotic dissociation from their own bodies: 'After a while, in fact, we no longer feel anything. It's someone else who is hurt, someone else who gets burnt and feels pain.'[15] The association between twins and animals is reinforced through references to anthropological accounts. In *The Notebook*, the twins enter into a strange friendship with the daughter of a neighbour who is in a predicament that mirrors their own. Like the boys, the neighbour has been forced by extreme poverty into an embrace of amoral survival tactics (theft) and animality (the twins spy on her having sex with their grandmother's dog). Due to her most visible deformation, she is known as 'Harelip,' a condition that is often associated in folk tales with both animality and twinship. In *The View from Afar*, Claude Lévi-Strauss points out that in Amerindian thought, a harelip is viewed as the 'beginning of a body splitting in two'. As a consequence, 'through this anatomical particularity, the Hare appears as a potential pair of twins'.[16] Through this reference, the neighbour is presented as a potential pair of twins in a way that seemingly pre-empts a suggestion evoked in the final volume of the trilogy that the twins were really one person all along and that twinship was an illusion conjured to endure the solitude of abandonment. Furthermore, the violent dissociation connected with twinship in the novel, it becomes clear, is a symptom not only of their abandonment but also of the trauma inflicted on the land itself by the war, a land with which they have become co-extensive. Throughout *The Notebook*, the violence inflicted on the Earth through bombing and neglect is presented as a corollary of the physical and psychological abuse inflicted on the children in the village.

The unnamed twins of Kristóf's *The Notebook* epitomize the use of twinship as a model of anthroposcenic performativity. Twinship is a strategy of survival of crisis through an erosion of individuality and an embrace of entanglement. Although it is not explicitly a story 'about' climate change, the novel inhabits what Mark Bould terms the 'anthropocene unconscious' by narrating a series of connected transformations: the destruction of the land through industrialized warfare and the transformation of human subjectivity in response to this destruction.[17] Rather than a denunciation of environmental catastrophe, Kristóf's twins constitute a myth commensurate to the Anthropocene and a guide to its survival. This chapter explores three dimensions of the role of twins in the myths of climate crisis. The first is the figure of 'impossible twins' developed by Brazilian anthropologist Eduardo Viveiros de Castro in his account of Amerindian 'perspectivalism'. The second is the use of twins to conceptualize non-anthropocentric alternatives to the dominant temporalities of modernity. The third is the use of twins as a conceptual tool for the 'scale-shifting' required by critical responses to the climate crisis, a lens through which to envision connections between the human, the micro-scale of cellular structures and the macro-scale of planetary processes. Twinship is mobilized in constructing an imaginary of what Kathryn Yusoff describes as 'an environmental relation to come', one that does not take as its foundation the myth of liberal individuation.[18] All of these anthroposcenic performances of twinship involve a paradox. While they employ twinship as a model for thinking beyond anthropocentrism they do so by instrumentalizing twins in a way that unconsciously repeats the very modern hierarchies of what Danowski and Viveiros de Castro describe as the 'anthropo-engineering' myth.

'Impossible Twins'

In *The Ends of the World*, Danowski and Viveiros de Castro propose a mythology that is 'adequate to our times', which they derive not from global popular culture or European philosophical traditions. Rather, the possible ontological solution they outline is drawn from the encounter between Western anthropology (particularly the brand of introspective self-erasing anthropological discourse that developed through the 1990s and 2000s in response to post-structuralism) and Amerindian cosmology. In this 'mythocosmological variant', rather than the 'world' being subtracted from its correlation with the human at the *end* of times (as in the techno-utopian discourse of human exceptionalism in which the human triumphantly transcends its material context marking the apex of modern progress), the disconnection is located at the *beginning* of times. 'The human is placed as *empirically anterior* in relation to the world.'[19] This schema is derived from Viveiros de Castro's previous work on Amerindian thought for which, he claimed, humanity or personhood, rather than being the culmination of a process of civilized socialization, 'is both the seed and the primordial ground, or background, of the world'.[20] In this set-up, the 'whole world' is included within the category of an amorphous and dynamic 'proto-humankind'.[21] The dimension of this cosmology that is most counter-intuitive to Western modernity is that '[i]t is Nature that is born out of or "separates" itself from Culture, not the other way around'.[22]

This concept is at the centre of what Viveiros de Castro describes as 'Amerindian Perspectivalism' in his 2009 single-authored book *Cannibal Metaphysics*. In Amerindian thought, Viveiros de Castro claims, the category of the human is a question of perspective. But rather than 'Nature' being the stable perspectival anchoring point for a multiplicity of 'Cultures' (as per the neoliberal discourse of multiculturalism), it is the other way around. A common 'Culture' gives way on to a multiplicity of 'Natures' (hence his neologism 'Multinaturism'). This 'Culture', however, rather than being a reified essential entity is characterized as a dynamic 'reservoir of difference', that holds 'the infinite potential for transformability'.[23] As he warns us: 'When everything is human, the human becomes a wholly other thing'.[24] The relevance of this cosmological vision to debates concerning the changing conceptions of nature-culture relations required to confront the current predicament is clear. Since the Amerindian populations of South America are the survivors of the genocide that was foundational to European modernity, this mythology is particularly valuable in the context of the climate crisis. As McKenzie Wark puts it, 'since they have already survived endings of the world, they may be better equipped for the Anthropocene'.[25]

In *Cannibal Metaphysics*, Viveiros de Castro uses twins as a model for the concept of dynamic difference that is at the heart of what it means to be human within Amerindian thought. In Amerindian cosmologies, he claims, twins are used as a model not of sameness or symmetry but of the 'dynamic difference' and 'perpetual disequilibrium' that are the driving motors of life.[26] Amerindian thought embraces the idea that difference 'only blooms to its full conceptual power when it becomes as slight as can be: like the difference between twins'.[27] Identical twinship is 'impossible' within this cosmology as sameness results in death. This account of twins within Viveiros de Castro's philosophy emerges from the meeting point between the twins or quasi-twins that feature in Amerindian myths and the central role played by the figure of twins within Claude Lévi-Strauss's work. Although myths about twins held a strong interest for Lévi-Strauss throughout his career, it was only in the analysis of Amerindian myth that he developed in the late books of the 1980s and 1990s that twins became in his eyes the 'key to the whole system', the 'password of Amerindian mythology and sociology'.[28] The account of twins he develops in *The Story of Lynx* is, in turn, key to Viveiros de Castro's surprising re-reading of Lévi-Strauss, the godfather of structural anthropology, as a precursor of what he describes as a poststructuralist 'anthropology of immanence'.[29] The 'dynamic disequilibrium' that characterizes the impossible twins in this work (first published in 1991) provides the keystone of the dialogue between Amerindian thought and Gilles Deleuze's philosophy of difference that the Brazilian Anthropologist brokers in *Cannibal Metaphysics*.

Viveiros de Castro's use of twins differs from their role in the dominant traditions of Western anthropology in the twentieth century. The assumption behind the focus on twins in the work of E.E. Evans-Pritchard and Victor Turner, for instance, is that the disruption of social conventions introduced by the 'anomaly' of twins has the side-effect of revealing the logic of the social structures that are being undermined. As William Viney puts it, the special treatment of twins, from celebratory ritual to violent exclusion, is 'an expression of a collective desire to restore an ideal order' and therefore 'attempts to incorporate and manage difficulties associated with twins helps to clarify ideals that are otherwise hidden.'[30] Whereas the dominant anthropological traditions of the twentieth century used twins as what Viney terms 'equipmental figures' with which to categorize kinship systems and religious practices, for Viveiros de Castro the twins of Amerindian myth embody the perpetual undoing of classification through a recursive dynamic of difference. Provocatively, he finds the model of this form of unstable twinship in the work of a figure often taken to be the embodiment of hegemonic Western anthropology. It is this central, instrumental, use of twins by Viveiros de Castro that highlights the fact that what David Chandler and Julian Reid describe as the 'ontopolitical turn' in anthropology entails 'a new form of colonisation of indigeneity for Western consumption'.[31] In a way that echoes the reduction of twins to scientific instruments in genetics research, indigenous cultures are made to embody the preoccupations of Western philosophy in a way that is, in the words of Alcida Rita Ramos, 'indifferent to the historical and political predicament of indigenous life in the modern world'.[32]

As a starting point of a global theory of twin myths that he starts to develop in *The Story of Lynx*, Lévi-Strauss establishes a distinction between Old World and New World twins. Whereas the former are used to conceptualize the production of unity out of the difference, the latter become the embodiment of inequality, perpetual instability and disequilibrium. Lévi-Strauss takes as his starting point a prior distinction between two different mythological 'formulas' for treating twins.[33] An example of the first formula are narratives about twins of opposite sexes who are 'doomed to an incest already prefigured by their promiscuousness in the maternal womb'.[34] This twin incest 'schema' stages the emergence of duality out of unity, albeit the 'unity' is equipped with built-in ambiguity. In the second formula, narratives about twins of the same sex are employed to perform the inverse dynamic: the emergence of unity out of duality. 'Can duality reabsorb itself into the image of the unity through which it is represented, or is it irreversible to the point that the minimal gap between its terms must fatally become greater?'[35] Mythical 'solutions' to this question range across the spectrum from one extreme in which twins embody irreducible difference (good twin, bad twin, etc.) to another in which they stand in for the transcendence of unity out of difference (the sharing of immortality and mortality by the Dioscuri, for example). While examples of the first schema can be found in the Dogon cosmology and mythologies of Southeast Asian, the dominant twin myths in Europe and the Americas are variations on the second formula. Whereas the Old World 'favors extreme solutions' of twins who are either identical or antithetical, American myths provide more 'graduated solutions' of 'relative inequality'.[36] The impossible twins of the Americas embody a 'dualism in a perpetual state of disequilibrium'.[37]

Coyote and Lynx, the protagonists of a tale told by the Nez Perce people of the Northwest coast of North America, are exemplary impossible twins. The story centres on a rivalry between these sworn adversaries, which culminates with Lynx covering Coyote's settlement with a thick fog that he only shifts when he becomes the new leader of his enemy's followers. The first chapter of *The Story of Lynx* traces variations on this tale across the Americas that are linked by a common attitude to opposition. For the relationship between Coyote and Lynx is 'akin to twinness': 'Perhaps alike originally – that is, twins from an anatomical perspective – the animals decided to become differentiated: Lynx lengthened Coyote's nose and legs, while Coyote pushed in Lynx's nose and shortened his tail.'[38] The likeness between the two is the temporary starting point of their transformation in opposite directions. The changes undergone by each, meanwhile, are conditioned by that of the other in a series of interlocking feedback loops. Lévi-Strauss argues that these impossible twins are exemplary of the form of social dualism that predominates across the Americas: a 'dualism in perpetual disequilibrium, whose successive states are embedded into one another'.[39]

Viveiros de Castro identifies a correlation between Lévi-Strauss's concept of impossible twins and the work of Deleuze and Guattari on two levels. In a way that echoes Lévi-Strauss's account of Amerindian thought, the philosophical system articulated in the 'Capitalism and Schizophrenia' books is populated by unstable dualisms. The construction of dualities is a key strategy in both *Anti-Oedipus* and

A Thousand Plateaus, from the opposition between arborescent and rhizomatic to the molar-molecular dyad. However, like the inequalities between impossible twins, the distinctions separating these dualities are never stable or untroubled. Rather, they are 'constantly interrupted by provisos, qualifications, involutions, subdivisions and other argumentative displacements'.[40] In both Deleuze and Guattari's dyads and the impossible twins, binaries are paradoxically used to undo dualistic thinking. The second correlation that Viveiros de Castro identifies between Lévi-Strauss's impossible twins and Deleuze and Guattari is their shared conception of relationality. Both systems presuppose a primacy of relation over identity. The development of singular divergences between Coyote and Lynx is conditioned by their interaction with one another in a way that echoes the actualization of singularities out of a multiplicity of virtual relations. However, in both cases relation is predicated on separation. The form of relationality that operates between impossible twins is what Deleuze calls 'disjunctive synthesis': a 'relational mode that does not have similarity or identity as its (formal or final) cause, but divergence or distance'.[41]

A particularly vivid variant on the impossible twins motif can be found in *The Old Capital*, a novel published by Japanese writer Yasunari Kawabata in 1962. The cultural and philosophical contexts of the book are very different to those of the mythological system traced in *The Story of Lynx* and *Cannibal Metaphysics*. However, the use of twins by Kawabata in the performance of a creative disjunction strongly echoes Lévi-Strauss and Viveiros de Castro's Amerindian twins. Moreover, the use of twins in Kawabata to interrogate the connection between technology and nature is illustrative of the power of impossible twins as a mythotechnological schema for the Anthropocene. *The Old Capital* tells the story of twin sisters born in Kitayama, a cedar-logging village in the mountains to the north of Kyoto, which was the home of the Emperor of Japan until the seat of government moved to Tokyo in 1868 as the country 'opened' to trade with the west. Driven by a mixture of economic necessity and a traditional fear of twins, one of them is abandoned and brought up by the owner of a traditional obi weaving workshop in the centre of the city. Chieko grows up in a Kyoto in which the world of her parents – a world dominated by aesthetic contemplation and the rhythms of the regular Shinto-Buddhist festivals – is encountering the encroachments of capitalist modernity that is having an increasing impact since the postwar occupation, even in the sheltered world of the 'old capital'. Traditional kimono retailers are selling radios alongside their fabrics and smaller businesses are being forced to change their business models to remain competitive. When Chieko visits Kitayama to view the cedars (the straight trunks of which 'make her spirit feel refreshed'), she encounters a girl of her age who seems to be almost identical in appearance.[42] When they meet again by chance at the annual Gion festival held in the city centre, it emerges that the girl is her twin sister Naeko.

The narrative at first seems to be following the model of what Lévi-Strauss described as an 'extreme solution' to the question of whether unity can emerge out of duality. While Chieko, the daughter of relatively wealthy artisans, wears a kimono chosen to establish the most pleasing contrast with the sombre hues of the cedar forest, Naeko wears the simple navy blue outfit of a fieldworker. Everything seems set for a variation on the 'prince and the pauper' narrative in which separated identical twins are used to interrogate the changing role of class and privilege in post-war Japan. This expectation is quickly thwarted as the account of the developing relationship between Chieko and Naeko emphasizes an immediate sense of familiarity. During their first encounter, Chieko 'felt a warm closeness to Naeko that filled her heart'.[43] However, the apparent harmony produced by the reunion between the two twins sets up a series of expectations that are also thwarted. Critics have famously described Kawabata's construction of a highly aestheticized fictional world stripped of all but the most minimal references to historical contexts, which privileges a concept of beauty that is premised on purity as being 'collusive with fascist aesthetics'.[44] The peaceful 'closeness' produced by the meeting of Chieko and Naeko at first seems to fit this model. Set against the backdrop of an unsettling modernization, the reunion of the twins seems to be a metaphor for a 'return' to nature and tradition.

The expectation that a seamless unity will emerge out of the duality of separated twins, healing the ruptures of an imposed modernity, is displaced by the novel in several ways. In the last section of *The Old Capital*, Naeko accepts Chieko's invitation to stay the night in her family home above the obi workshop. The image of the two girls sleeping together in the same room is described in terms that evoke a return to the womb, enclosed and protected from the winter elements. The vision of them huddled under their bedclothes echoes a previous sequence when, caught in a sudden downpour in the mountains around Kitayama, Naeko shelters Chieko from the rain. Thanking her sister, Chieko exclaims: 'I wonder if you did this for me in our mother's womb.'[45] But just as the reader thinks that this might be the beginning of an enduring sisterly relationship, Naeko appears to carry out her threat to 'disappear completely' when she leaves the house into the snow without looking back, having refused her sister's offer of an umbrella and coat. This jarring ending is a variation on a strategy used in the denouements of a number of Kawabata's novels including *Snow Country* and *A Thousand Cranes* in which an abrupt transition violently undermines the narrative edifice that had been constructed up until that point in the text.

Carl Cassegard describes this dynamic that is characteristic of Kawabata's fiction as a 'shock'. He opposes this to Walter Benjamin's account of 'shock' as a reaction to the erosion of 'experience' by the fragmentations of modernity. Shock, in this case, is constitutive of the 'shield of identity' constructed by human subjects to protect their 'inner nature' against the upheavals of the capitalist imperative of constant change and transformation ('all that is solid melts into air'). In Kawabata, by contrast, shock produces a disintegration of identity and an encounter with 'nothingness' that is potentially liberating rather than imprisoning. Nothingness, Cassegard points out, is the dynamic nothingness that Kawabata associated with the Taoist concept of 'mumei' in his Nobel Prize acceptance speech of 1968: 'This is not the nothingness or the emptiness of the West. It is rather the reverse, a universe of the spirit in which everything communicates freely with everything, transcending bounds, limitless.'[46] From this perspective, as Cassegard goes on to explain, shock is 'the fissure in identity which admits the light of the limitless, unrealized possibilities that were once open. Shock wants to return us to the state before any of our particularities took form.'[47]

In the ending of *The Old Capital*, twinship is employed as an agent of shock in a way that echoes the impossible twins of Amerindian cosmologies. Twins are never fully aligned with tradition in opposition to modernity in the novel. At one stage, Chieko's adoptive father welcomes Naeko into their home explaining that '[t]wenty years ago twins weren't accepted, but now it's nothing'.[48] The implication is that the rapid globalization of the post-war period has swept away the folk superstitions lingering in rural communities. The replication of twins in this case is used as a metaphor for the industrialized production methods that are replacing the artisanal methods practised by the fabric weavers of Kyoto. Furthermore, despite initially seeking her sister out, Naeko grows increasingly uncomfortable with the twin bond, describing it in cryptic terms as an 'illusion'. Although never explained in uncertain terms, it becomes clear that what she means by this is that their twinness has become an object of the form of aesthetic contemplation known as *mono no aware*. As they contemplate themselves through the new perspective produced by the knowledge of their sisterhood, their separation as subjects is momentarily interrupted in a way that evokes the disruption of melancholy. If their twinness holds the promise of a return to a lost unity, it does so only briefly and, as Naeko is aware, the illusion will soon be broken.

The disjunctive dynamic that develops between Chieko and Naeko functions as a metaphor for entanglements between nature and technology in the novel. *The Old Capital* presents technology and nature as impossible twins, locked together in a disjunctive synthesis. At the time of their first encounter, Chieko is associated with the sophisticated weaving techniques practised by her family, while Naeko is associated with the natural abundance of the Kitayama area. However, the separations between the two are constantly undermined. The nature of Kitayama is a very visible artificial nature. The cedar trees that grow there were planted to be used in construction and for ceremonial purposes and this fate seems to haunt their presence on the hillside. 'The cedars were used in the construction of tearooms, so the groves themselves had the elegant air of the tea ceremony.'[49] The artisanal techniques employed by Chieko's family and their milieu, meanwhile, are presented as having been inherited, a natural trait passed down through the generations. The designs produced for the obis are developed through contemplation of nature, as if the designer's job is merely to channel the beauty already existing in the natural world. Cassegard argues that shock in Kawabata's work functions as a 'road back to an originary state of nature.'[50] While twinship is an agent of shock in *The Old Capital*, it produces not a 'return to nature' but a violent encounter with a techno-nature that, far from being elegiac, is dynamic and transformative.

The echoes of the cybernetic twins discussed in Chapter 1 are clear. The developmental paths taken by impossible twins are dictated by what Yuk Hui identifies as the dual cybernetic principles of recursion and contingency. Recursion, Hui argues, 'is not mere mechanical repetition; it is characterized by the looping movement of returning to itself in order to determine itself, while every movement is open to contingency, which in turn determines singularity.'[51] Rather than uniting Chieko and Naeko in a unified being, the recursive dynamic of exchange that defines the relationship between the two sisters sublates the opposition between being and becoming: 'Being is preserved as a dynamic structure whose operation is open to the incoming of contingency: namely, becoming.'[52] The model of techno-natural 'being' provided by the twins in Kawabata is characterized by what Hui labels a 'mechanical organicism' of cybernetic recursion. Furthermore, the dynamic that unfolds between self and other in Viveiros de Castro's perspectivalism closely echoes that of the enemy within cybernetics. Consciousness about the self is reached, not through confrontation with an Other, but by occupying the Other's perspective temporarily. In Viveiros de Castro's interpretation of Tupinambá cannibalism, the victim and executor are thought of as twins. Not only would the executor not eat the captive, but afterward he would also enter into a period of mourning expressed through a funerary confinement. He engaged in 'a process of identification with this "opponent" whose life he had just taken'.[53]

However, if impossible twins offer a model of relationality for the Anthropocene it is a mode of relation that is structured by disjunction. Frédéric Neyrat argues that while ecological thought should oppose the modern separation of man and nature that has resulted in an instrumentalization of the Earth, those seeking critical modes of being in the Anthropocene should be equally as suspicious of the idea that 'everything is interconnected'.[54] This 'principle of principles' of ecology and environmentalism has been an important critical tool in the fight against 'denial of relation' on which 'the modern era laid its foundation'.[55] However, the full embrace of interconnection also poses a problem for critical thought since the absence of a distance forecloses the possibility of change since both humanity and world are caught up in the same 'ontological turbulence'.[56] Against the two extremes of modern separation and pervasive interconnection, Neyrat proposes an 'ecology of separation' that might 'open the Earth up to its own outsides'.[57] The disjunctive synthesis of impossible twinship provides a model, at a human scale, of the interior distance needed to think and live this form of ecology. As Viveiros de Castro points out, the bond of twinship does not erase divisions. Rather, it 'undefines' them by 'bending every line of division into an infinitely complex curve'. Impossible twinship 'is not a question of erasing the contours but of folding and thickening them, diffracting and rendering them iridescent'.[58]

Weird temporalities

Twinship has a complex relationship with time. In his account of twin research, Lawrence Wright describes the events that lead to the spectrum of monozygotic twinship, including conjoined twins and *fetus-in-fetu*, as 'accidents of timing during early pregnancy'.[59] Twins that share a placenta result from disruptions of the usual chronology of single embryonic development. If the zygote splits into two embryos within the first three days of its existence it is likely to be a dizygotic pregnancy leading to what are popularly known as 'fraternal' twins. If the split takes place after that point then the pregnancy is likely to be monozygotic, leading to the misleadingly named 'identical' twins. A divide that takes place after around the thirteenth day can result in various degrees and forms of conjoinment. Timing is everything. This awkward relationship with time during the early stages of pregnancy accompanies twinship after birth. In her description of biocultural existence, Samantha Frost emphasizes the importance of time in our considerations of the porosity of boundaries between human organisms and their environments, outlined in the Introduction. In the postgenomic era, she argues, time has become a crucial factor in our understanding of the genome and how genetic information conditions human development. Drawing on the work of Martine Lappé and Hannah Landecker, Frost points out that the genome is now thought of as having a 'varied temporal horizon' or 'life span' within which 'changes in environmental provocations are registered at the level of chromatin'.[60] Biological processes do not pre-exist environments (this would be biological essentialism) but are *constituted through* responsiveness to those environments'.[61] However, this responsiveness is characterized by a time lag since the body's responses (its ongoing activities of 'composing and recomposing') are 'noncontemporaneous' with the environmental provocations of its current habitats. The speed of reaction can vary enormously as the body can encode responses to both its own lived environments and to the environments of previous generations.

The complex interplay between sameness and difference in monozygotic twins exposes the genome's 'varied temporal horizon'. While sameness can speak to bodily responsiveness to the environments of previous generations, differences reveal more immediate bodily responses, from the different experience of touch during the womb (which explains why monozygotic twins have different fingerprints) to the accidents and interventions of bodily scarring and manipulation. If twins are employed in genetics research as media through which to articulate the connection between genetic information and identity, then this connection is characterized by a temporality that is extremely complex. When you look at the faces of 'identical' twins the eye tends to scan back and forth in search of differences. These visual differences index the epigenetic encoding of the series of environmental encounters starting at the moment of zygotic splitting. The search back and forth across faces is a search back and forth across temporal scales. In the process, they confront us with the multi-scaler temporalities of the human organism's entanglements with its environments. The confrontation with this varied temporal horizon is at the centre of the fascination exerted by photographic portraits of identical twins. Whereas Galton's 'Twin Method' inaugurated a photographic tradition in which twins were seen as a natural embodiment of technological reproduction and sameness, there is a strong counter-tradition that emphasizes this complex interplay between sameness and difference and presents the replication of twins as a form of diffraction. The subtle asymmetry of the iconic photograph of the 'Identical Twins, Roselle, New Jersey' discussed in Chapter 5 is indicative of a parallel that runs through Diane Arbus's work between photographic and biological reproduction as forms of unstable and productive mutation.

The concept of the Anthropocene poses radical challenges to the concepts of time that structure the modern worldview. The myth of progress that frames time in terms of a linear one-way developmental path that culminates in the triumph of modernity has been fundamentally shaken by the climate crisis. One of the dominant narratives in environmental thought presents the crisis as a revenge enacted by nature on the excesses of modernity. From this perspective, violence enacted on the Earth at the birth of the modern world loops back to haunt humanity's collective future. The concept of the Anthropocene merges two temporal scales that modern thought has kept strictly separate: the scale of modern social history, measured out in human lifetimes, political regimes or cultural systems, and the scale of geological time, measured in millennia and lithic strata. Timothy Clark argues that we are living through a 'widespread crisis of scale' that poses drastic challenges to the concept of human agency that undergirds Western liberalism.[62] While an inability to think across the scales of the individual and the planetary has partly caused the crisis, it is now one of the toughest challenges facing ecological activism. As Staicy Alaimo puts it, 'the ethical and the political, like many other questions of and in the Anthropocene, become matters of scale-shifting – improvisational interventions in lives and worlds where there is no stable background and nothing can be set straight.'[63]

Twins have long been used to think across scales. Eugenics uses twins to impose the smooth scalability of modern industrial production onto human life. As a consequence of this use, monozygotic twins have come to embody this modern concept of scale. As Anna Tsing describes it, '[s]elf-replicating things are models of the kind of nature that technical prowess can control: they are modern. They are interchangeable with each other, because their variability is contained by their self-creation.'[64] In their use in genetics research, twins become the embodiment of a powerful discourse that, in Tsing's words, 'captured genetic inheritance in a self-enclosed and self-replicating modernity'.[65] However, the representational tradition embodied by Arbus's 'Identical Twins, Roselle, New Jersey' uses twins to enact a form of 'scale shifting' radically opposed to modern industrial scalability, which is closer to that demanded by Alaimo. Confronting twin faces requires an act of temporal scale-shifting. The interplay between sameness and difference in monozygotic twins forces us to think across temporal scales, between the intergeneration time scales of genetic inheritance and the most immediate temporality of lived experience. If behavioural geneticists use twinship to forge stable connections between different scales – the scalability that is central to modern systems – then these photographic uses of twins evoke a complex assemblage of different temporal scales, subject to what Tsing describes as 'indeterminacies of encounter'.[66] So, while twins become a lens for scale-shifting, they themselves are presented as non-scalable: a blockage both in the temporality of progress and modern systems reliant on smooth machinic scalability.

The forms of temporality and scale enacted by twins are weird in the sense traced in Chapter 2. In his study of weird aesthetics, Mark Fisher draws attention to the etymological association between the word 'weird' and the concept of fate. The Oxford English Dictionary describes the adjective 'weird' as 'having the power to control the fate or destiny of human beings, etc.; later, claiming power of dealing with fate or destiny'. The naming of the sooth-saying witches in *Macbeth* 'Weird Sisters' is probably the most familiar example of this meaning of the word. Fisher explains that the concept of fate is weird since 'it implies twisted forms of time and causality that are alien to ordinary perception'.[67] But, to use another term developed by Fisher, fate is also 'eerie' in that it 'raises questions about agency: who or what is the entity that has woven fate?'[68] Weird temporalities are often evoked in popular culture through the construction of what Douglas Hofstadter has described as 'strange loops' of 'tangled hierarchies' that disrupt neat distinctions between cause and effect.[69]

In *Dark Ecology*, Timothy Morton argues that one of the challenges of the current crisis is the requirement it confronts us with to think 'at temporal and spatial scales that are unfamiliar, even monstrously gigantic'.[70] Rather than simply think at a global scale, the challenge is to think across vastly divergent scales simultaneously. Ecognosis, the mode of thinking proposed by Morton, should be alert to 'dark ecological loops' that connect different scales in unintended and unexpected ways. One example of a dark ecological loop cited by Morton is the decimation of bee populations at the beginning of the century caused by pesticides. 'Such unintended consequences are weirdly weird in the sense that they are uncanny, unexpected fallout from the myth of progress: for every forward motion of the drill bit there is a backward gyration, an asymmetrical contrary motion.'[71] The temporality of the Anthropocene is weird in that it loops together 'two levels we usually think are distinct: geology and humanity'.[72]

The trope of the double is often used to construct these strange loops. One of the examples cited by Fisher is the ending to Stanley Kubrick's *The Shining* (1980) in which, while the Overlook Hotel's new caretaker Jack Torrance (played by Jack Nicholson) is trying to murder his family in the temporal present of the 1970s, the viewer is shown an image of Torrance in a photograph of the hotel summer ball in 1921. This twist loops the plot back to the beginning by revealing that Jack has always been the caretaker of the Overlook and that his fate was inevitable. This 'twisted causality', Fisher explains, produces a charge of the weird.[73] Lynch's *Lost Highway* which, as discussed in Chapter 2, carries out a moment of slippage between twinship and doubleness, also uses the trope of the double to construct a weird, looping temporality. The film opens with Fred Madison receiving a message on his intercom that 'Dick Laurent is dead'. The ending reveals that the message was left by none other than Fred himself who, due to a strange loop in time, is trying to warn himself of an impending crisis.[74]

In the studies of twins raised apart emerging from the MISTRA project (discussed in the Introduction), the temporality evoked by twins is weird in the sense outlined by Fisher. The meeting of James Alan Lewis and James Allen Springer that is used as the project's origin story has a similar narrative structure to *The Shining*. The encounter between the two separated brothers induces a temporal loop that reveals that their fates are not the product of expected factors such as life experience and childhood environment but are largely the result of their genetic code. As 'viewers' of this research, the meeting of estranged twins such as the Springer brothers has the same effect as the black-and-white photograph at the end of Kubrick's film. In *The Shining*, the viewer is given retrospective knowledge that the tension produced by Jack's conversion into the murderous caretaker was unnecessary since he was the caretaker all along. Similarly, the desired effect of Nancy Segal's account of the Springer twins in *Born Together – Reared Apart* is to reveal the drama of life experiences and childhood environment as being almost inconsequential seeing as genes were the crucial determining factor from the beginning. The use of twins in behavioural genetics research produces a looping temporality that those used to crediting environmental factors such as cultural context with a determining power reject as weird. A sense of futility contributes to the weird affective charge produced by time travel narratives like *12 Monkeys*. Both critics and admirers of the vision of personalized genetics outlined by Robert Plomin in *Blueprint* give expression to a similar sense of futility: more often than not our current one-size-fits-all education systems are at odds with our children's genetically determined predispositions and capabilities.[75]

In the context of climate crisis, the medium of twinship has been used to estrange us from the temporalities of modernity that have brought us to the brink of catastrophe in the first place. This is clearest in Shelley Jackson's 2006 novel *Half Life*, the protagonists and narrators of which are conjoined twins.

In temporal, thematic and structural terms, the book sits half-way between Jackson's 1995 work of hypertext fiction, *Patchwork Girl*, for which she is still best known, and her 2018 postmodern gothic novel *Riddance: Or: The Sybil Joines Vocational School for Ghost Speakers and Hearing-Mouth Children*. *Half Life* revisits some of the key themes surrounding twinship, including its challenge to the concepts of human subjectivity as separate and unique, through the conventions of the gothic double and the structural strategies associated with North American metafiction of the latter half of the twentieth century, primarily the disruption of the plot through self-conscious and often obtrusive references to both intertexts and narrative technique. From *Patchwork Girl*, the novel borrows both its poststructuralist-feminist critical stance (Avital Ronell and Judith Butler are both mentioned in the acknowledgements) and its conception of narrative as a game that only a highly active and critical reader can enter into. Read alongside Jackson's companion piece essay 'Stitch Bitch' (1997), *Patchwork Girl* was both a work of fiction to be enjoyed and a theoretical framework for the then nascent form of the hypertext novel. Similarly, *Half Life* questions and undermines the theoretical premises of its own media, both the medium of the codex and that of twinship. The novel employs conjoined twinship as a medium through which to articulate a challenge to the temporalities associated with Western modernity's dominant print cultures. The weird temporalities of twinship that result from this challenge are presented as responses to the ecological crises that are the culmination of Western modernity, its own repressed double returned to haunt it.

The novel is narrated from the perspective of Nora Olney, whose twin sister Blanche, with whom she is a conjoined 'standard issue twofer' (two heads, two legs and two arms), has been asleep for 15 years.[76] The narrative trigger is Nora's decision to have her sister's head removed so that she can become the 'singleton' she has always dreamed of being. As Nora attempts to track down an underground semi-mythical organization that can carry out this operation, there are signs that Blanche is starting to wake up, including uncontrollable activity by the arm on her side of the body (mainly throwing objects across the room) and the fact that Nora keeps experiencing hallucinations as if, she suspects, her sister is gradually encroaching on her consciousness. The narrative is caught within a tension between Nora's quest for separation, and her investigation into the nature of her twin connection. This investigation involves both tracing Blanche's suspected activities in the narrative present (it emerges that Blanche might simultaneously be orchestrating the removal of Nora's head) and an imaginary return to their shared childhood in the deserts of Nevada. For, Nora gradually realizes, to come to terms with Blanche's constitutive role in her existence is also to come to terms with her own repressed past. The structure of the novel mirrors this tension. Chapters narrating Nora's quest and investigation in the present alternate with chapters recounting episodes from their childhood. Both are interspersed with excerpts from Nora's 'Siamese Twin Reference Manual', a scrapbook of cuttings and information about the 'twofer' phenomenon. For Nora and Blanche's conjoined existence is not as rare in the world of the novel as it would be for the reader. As a newspaper clipping from the *San Francisco Herald* tells us, the novel takes place in a parallel universe in which, in an act of collective atonement for dropping the nuclear bombs on Hiroshima and Nagasaki, the United States detonated a series of bombs on its own territory in a 'National Penitence Ground' located in the deserts of Nevada. Though there were competing theories, most believe that the 'dramatic growth in the conjoined twin population' was caused by radioactive fallout.[77]

The main two temporal frameworks governing the novel are that of Nora's investigation, constructed through conventions borrowed from the detective genre, and that of her attempts to come to terms with the past and incorporate her repressed memories into her lived present (through the narrative conventions of mourning). The temporalities of both detection and mourning are undermined. Firstly, the ending of the novel frustrates the expectations that have been set up by Nora's presentation of herself as a detective following clues left by a criminal, her sister Blanche. The way that the narrative about the twins' childhood gradually catches up with the present seems to be leading to a revelation about the nature and extent of Blanche's influence over Nora. However, this expectation is frustrated by the abrupt 'unhappening' that takes place during the final pages of the novel.[78] Rather than resolution, the reader is confronted with a violent implosion that undoes the fragile distinctions that have been sustaining the narrative through its 437 pages. As Amanda Greer puts it in her assessment of *Half Life* as an antidetective 'Gumshoe Gothic' story, '[i]n one body Nora and Blanche gather together criminal, detective, and victim, and distinguishing between them is as difficult as distinguishing where Nora ends and Blanche begins'.[79] The final words of the novel – '"Nora?" I say' – is the culmination of a suspicion that has been building throughout the novel.[80] Nora and Blanche are as opposite as black (noir) and white (blanc). But they are co-constitutive just as black type only acquires meaning in opposition to the white of the page. During the implosive ending, the narrative acknowledges this in a moment of apocalyptic self-knowledge. 'I have spent my whole life trying to make one story out of two: my word against Blanche's. But we are only as antithetical as this ink and this page. Do these letters have meaning, or the space around them? Neither. It's their difference we read.'[81] The suggestion that it is

Blanche writing the final words collapses the distinction between the two sisters in a way that echoes William Wilson's suicide at the end of Poe's paradigmatic gothic double narrative. In place of resolving the past into the present, in the manner of a traditional detective narrative, the ending leaves the reader with a strange temporal loop. By ending with the beginning of Blanche's account, the suggestion is that the two narratives could loop around endlessly in a way that evokes a moebius strip.[82] As Greer puts it, '[i]nstead of the expected revelation, the reader finds the world of the text flickering in indeterminacy'.[83]

The violence committed by the false ending to the distinctions that sustain the detective genre is mapped onto the politics of collective memory and mourning. The alternative historical context of the novel makes Nora and Blanche's internal struggles stand in for a nation divided. At one point, Nora compares the silences repressed from the narrative of her own life (embodied by Blanche) with the silences surrounding the nation's suppressed history of violence (embodied by the National Penitence Ground). The failed attempt by both Nora and the nation she embodies to properly mourn the past and incorporate it into a harmonious narrative of its identity in the present leads to melancholy. Nora and Blanche are melancholy twins in a way that echoes Harume and Akio in Fumiko Enchi's novel *Masks*, discussed in Chapter 2. Characterized by moebius strip-like loops, twinship in *Half Life* is a melancholy medium. The alternative historical backdrop of *Half Life* locates the weird looping temporalities it produces firmly within the context of the Anthropocene. In her study of ecological concerns in contemporary queer fiction, Nicole Seymour argues that, by actively exploring the new subject positions afforded by conjoinment, '*Half Life*'s citizens learn to simultaneously criticize the government's deployment of nuclear technology and embrace its fallout, for how it provides a different model for the future – one attuned to interdependence and interrelations among the human, the non-human, and those who fall in between.'[84] The flickering narration, and the looping temporality it produces, are, in Seymour's analysis, the stylistic manifestation of the queer ecology it articulates at a thematic level. Seymour's characterization of *Half Life* as a work of queer ecology points to the critical potential of the weird temporalities of twinship in the context of global climate crisis. In *Half Life*, twinship becomes a medium for this ecognostic perspective on weird temporalities. If behavioural geneticists use twinship to forge connections between different scales (molecular and molar), then Jackson uses her twins to create a loop between the distributed subjectivity of assemblages for which Nora/Blanche are presented as a model and a perspective of global interconnection and the planetary scale of climate crisis.

It is the flickering, melancholy structure of *Half Life* that the novel shares with Jackson's earlier work with electronic media. When the temporal trajectories in the novel implode, the reader is confronted with the brute materiality of the medium itself: the codex and its structural and typographical conventions. A full two-thirds of the final page, six lines above the cataclysmic last line ('"Nora?" I say.'), is left blank with the exception of a series of punctuation marks in imitation of Nora/ Blanche's hand-written journal that forms the basis of the novel itself. 'I look down and see that my ballpoint has been sponged clean by the wet paper. The page is blank, except for the occasional ding where I bore down on a comma or a dash.'[85] The repressed blanks of the paper – the absences that, together with the ink, are constitutive of meaning – are asserting their absent-presence just at the moment when Blanche reveals that she has been there and 'awake' all along. In his study of 'dark media' in the horror genre, Thacker argues that the 'endpoint of mediation is the negation of mediation itself'.[86] In the 'avant garde horror' *Outer Space* (1999), directed by Peter Tscherkassy, the logic of possessed, excessive mediation staged within the film, transgresses its fictional boundaries to infect the medium. '[B]oth film and sound overload the medium so completely that all we are left with is the actual, physical, material of the film itself.'[87] Rather than overloading the medium of the codex, in its culminating moment, *Half Life* strips it back to its skeleton. But the effect is the same: the hyper mediation of haunting spills over into a blocked, weird mediation and in the process reveals its own conditions of possibility and limitations.

Jackson's hypertext novel *Patchwork Girl* also sets out to reveal the potential and limitations of its own medium. David Punday argues that American metafiction of the 1960s and 1970s, by authors such as John Barth, provided a model for the game narratives favoured in electronic fiction of the 1990s. In both, a melancholy structuring aesthetic predominates. Through frequent recourse to intertextuality and metafiction, 'narrative is interrupted to such a degree that forward progress becomes impossible'.[88] In the process, the machinery of the textual systems set in motion are revealed as uncannily autonomous. Although Punday himself argues that *Patchwork Girl* is an exception to this rule since it presents the electronic medium as an 'instrument' that can be 'set at a distance and kept separate from the body of the reader and author,'[89] Katherine Hayles disagrees. In her interpretation, Hayles argues that *Patchwork Girl* exploits the fact that 'electronic hypertexts are written and read in distributed cognitive environments' and that the reader is 'constructed as a cyborg, spliced into an integrated circuit with one or more intelligent machines'.[90] By appropriating and re-writing a canonical eighteenth-century novel within this distributed textual ecology, Jackson is staging a return of the material and distributed dimensions of literary writing that were repressed in the articulation, by copyright law during this period, of literature as the emanation of individual masculine genius that 'soared above [its] material instantiations in books'.[91]

Jackson uses the specificities of her electronic medium – what she terms the 'flickering connectivities' that mediate between the author, reader and the multi-layered coding chains enabling the interface – to turn her text into the repressed unconscious of literary discourse. The patchwork girl of the title, a stitched assemblage of different bodies, stands in as a metaphor for the boundary-breaching textual performance required of the reader. If Punday argues that the print metafiction of the 1960s and 1970s informed the melancholy structural principles of electronic fiction of the 1990s, then Jackson reverses this by using her hypertext work as a model for the later print novel. In *Half Life*, Jackson uses the medium of the codex as a tool through which to expose the possibilities and limitations of digital media. On the one hand, twinship in the novel stands in both for the hypermediation and interconnection facilitated by network media. Nora and Blanche's conjoined subjectivity performs the distributed cyborg subjectivity of *Patchwork Girl*. On the other hand, by taking the recursive logic of metafiction to an extreme, they embody the moment when hypermediation spills over into the negation of mediation itself. Jackson uses print to expose the paradoxical logic of digital media (that hypermediation coincides with an absence of mediation) for the same reason that behavioural geneticists use twins to render legible the molecular workings of the body at a molar level. However, she does so to perform recursive feedback loops between the body and the environment, which cut through and across different scales of space and time.

Conclusion: Twinning as ecognosis

By reproducing the recursive logic of electronic fiction in a print novel about ecological crisis, *Half Life* draws attention to Hui's argument that 'modern machines and ecology are two discourses adhering to the same principle, namely, cybernetics'.[92] This points to an irony underlying the novel, which undermines Seymour's argument that the *Half Life*'s staging of interconnected subjectivities serves as a model for a post-anthropocentric future. The recursive dynamics of these subjectivities are grounded in a technological logic developed as part of a machinery of war alongside the atomic bomb that, within the novel, is the culmination of the technological modernity that has driven the ecological crisis in the first place. Hui argues that 'maybe it is no longer a dualism which is the source of danger in our epoch, but rather a non-dualistic totalizing power present in modern technology, which ironically resonates with the anti-dualistic ideology'.[93] So the recursive interconnected subjectivities staged within the novel as a radical alternative to liberal humanism embody the dominant cybernetic ethos of digital capitalism. By mirroring the logic of the system that it is critiquing, *Half Life* echoes the way that Viveiros de Castro repeats a colonizing gesture of appropriation during his attempt to place the perspectives of indigenous epistemologies on equal terms with Western systems of knowledge.

The form of ecognostic thought that twins embody in these texts incorporates these ambiguities into its moebius-like loops. The logic of modern technology returns as a model of radical ecological subjectivities, which in turn serves as a blue print for greater integration between humans and algorithmic systems. Twinning as a form of ecognosis echoes the concept of twinning as a model of 'methodological intimacy' discussed in the Introduction in one important respect. Twinning, Chiew and Barnweel explain, 'is a refusal to abstract figure from ground, or to pinpoint where one thing ends and another begins'.[94] Not only are twins models of 'intimate entanglement' in their relationships with each other, but the desire to explain twins also confronts us with an impossibility of separating the 'figure' of individuality from the ecological ground of genetic code and socio-cultural upbringing. Danowski and Viveiros de Castro point out that the concept of the Anthropocene performs a 'deadly inversion' of the relation between figure and ground since the 'ambiented becomes the ambient (or "ambienting")' and vice versa.[95] The Earth systems that have provided the passive and inert background for the promethean exploits of modern man have suddenly transformed into a 'menacing form of a historical subject' and 'political agent'.[96] This reversal violently undermines the myths of human exceptionalism that are premised on a strict separation between passive nature and active historical human subject. However, rather than a clean reversal of these binaries, which would relegate humanity to a position of hopeless passivity, the lens of twinning presents the connection between human and world to be recursive, obeying the anti-logic of the moebius strip.

Notes

1   James George Frazer, *The Golden Bough: A Study in Magic and Religion* (London: Wordsworth Reference, 1993 [1890]), 66.
2   Frazer, *The Golden Bough*, 66.

3   Frazer, *The Golden Bough*, 66.
4   Frazer, *The Golden Bough*, 67.
5   Déborah Danowski and Eduardo Viveiros De Castro, *The Ends of the World*, trans. Rodrigo Nunes (Cambridge: Polity Press, 2017), 62.
6   Danowski and Viveiros de Castro, *The Ends of the World*, 48.
7   Una Chaudhuri, 'Enduring Performance', Keynote Address at 'Art in the Anthropocene' conference, Trinity College, Dublin, 8 June 2019.
8   Yuk Hui, 'Machine and Ecology,' *Angelaki* 25:4 (2020): 57.
9   Donna Haraway, 'Symbiogenesis, Sympoiesis, and Art Science Activisms for Staying with the Trouble', in *Arts of Living on a Damaged Planet: Monsters of the Anthropocene*, ed. Anna Tsing, Heather Swanson, Elaine Gan and Nils Bubandt (Minneapolis: University of Minnesota Press, 2017), M39.
10 Ágota Kristóf, *The Notebook*, trans. Alan Sheridan (London: CB editions, 2014 [1986]), 4.
11 William Viney, 'The Significance of Twins in the Middle Ages', https://thewonderoftwins.wordpress.com/2013/07/23/the-significance-of-twins-in-medieval-and-early-modern-europe/.
12 Kristóf, *The Notebook*, 15.
13 Kristóf, *The Notebook*, 35.
14 Kristóf, *The Notebook*, 22.
15 Kristóf, *The Notebook*, 17.
16 Quoted in Claude Lévi-Strauss, *The Story of Lynx*, trans. Catherine Tihanyi (Chicago & London: The University of Chicago Press, 1995), 127.
17 Mark Bould, *The Anthropocene Unconscious: Climate Catastrophe in Contemporary Culture* (London: Verso, 2021)
18 Kathryn Yusoff, *A Billion Black Anthropocenes or None* (Minneapolis: The MIT Press, 2018), 17.
19 Danowski and Viveiros de Castro, *The Ends of the World*, 63.
20 Danowski and Viveiros de Castro, *The Ends of the World*, 65.
21 Danowski and Viveiros de Castro, *The Ends of the World*, 66.
22 Danowski and Viveiros de Castro, *The Ends of the World*, 67.
23 Danowski and Viveiros de Castro, *The Ends of the World*, 66–8.
24 Eduardo Viveiros De Castro, *Cannibal Metaphysics: For a Post-Structural Anthropology*, trans. and ed. Peter Skafish (Minneapolis, MN: Univocal, 2014), 63.
25 McKenzie Wark, 'Eduardo Viveiros de Castro: In and against the Human', *Verso blog*, 12 June 2017, https://www.versobooks.com/blogs/3265-eduardo-viveiros-de-castro-in-and-against-the-human.
26 Viveiros de Castro, *Cannibal Metaphysics*, 154.
27 Viveiros de Castro, *Cannibal Metaphysics*, 59.
28 Viveiros de Castro, *Cannibal Metaphysics*, 211.

29 Viveiros de Castro, *Cannibal Metaphysics*, 46.
30 William Viney, 'Anthropology's Twins', https://thewonderoftwins.wordpress.com/tag/levi-strauss/.
31 David Chandler and Julian Reid, 'Becoming Indigenous: The "Speculative Turn" in Anthropology and the (Re)colonisation of Indigeneity', *Postcolonial Studies* 23:4 (2020): 486.
32 Alcida Rita Ramos quoted in Chandler and Reid, 'Becoming Indigenous', 494.
33 Alessandra Piontelli points out that the twin myths that emphasize inequality over sameness are much closer to the intrauterine reality that became visible with the introduction of sonographic technology. 'Twins can be discordant with respect to potential risks and be affected in different ways. The intrauterine environment of twins can vary greatly: one twin can have a bigger placenta, less amniotic fluid, or an umbilical cord with just two vessels instead of three. [...] Paradoxically, so-called identical or monozygotic twins are more liable to suffer from intrauterine inequalities.' *Twins in the World: The Legends They Inspire and the Lives They Lead* (New York: Palgrave Macmillan, 2008), 69–70.
34 Lévi-Strauss, *The Story of Lynx*, 225.
35 Lévi-Strauss, *The Story of Lynx*, 226.
36 Lévi-Strauss, *The Story of Lynx*, 126–7.
37 Lévi-Strauss, *The Story of Lynx*, 235.
38 Lévi-Strauss, *The Story of Lynx*, 49.
39 Lévi-Strauss, *The Story of Lynx*, 239.
40 Viveiros de Castro, *Cannibal Metaphysics*, 116.
41 Viveiros de Castro, *Cannibal Metaphysics*, 112.
42 Yasunari Kawabata, *The Old Capital*, trans. J. Martin Holman (Berkeley: Counterpoint, 2006 [1962]), 63.
43 Kawabata, *The Old Capital*, 90.
44 Nina Cornyetz, 'Fascist Aesthetics and the Politics of Representation in Kawabata Yasunari', in *The Culture of Japanese Fascism*, ed. Alan Tansman (Durham, NC: Duke University Press, 2009), 348.
45 Kawabata, *The Old Capital*, 117.
46 Yasunari Kawabata, '"Japan, the Beautiful and Myself": Nobel Lecture', in *Dandelions*, trans. Michael Emmerich (London: Penguin, 2019 [1968]), 126.
47 Carl Cassegard, 'Shock and Modernity in Walter Benjamin and Kawabata Yasunari', *Japanese Studies* 19:3 (1999): 247.
48 Kawabata, *The Old Capital*, 164.
49 Kawabata, *The Old Capital*, 68.
50 Cassegard, 'Shock and Modernity', 247.
51 Yuk Hui, *Recursivity and Contingency* (Lanham, MD: Rowman & Littlefield, 2019), 17–18.
52 Hui, *Recursivity and Contingency*, 18.

53 Viveiros de Castro, *Cannibal Metaphysics*, 141.
54 Frédéric Neyrat, *The Unconstructable Earth: An Ecology of Separation* (New York: Fordham, 2018), 11.
55 Neyrat, *The Unconstructable Earth*, 11.
56 Neyrat, *The Unconstructable Earth*, 11.
57 Neyrat, *The Unconstructable Earth*, 15.
58 Viveiros de Castro, *Cannibal Metaphysics*, 44.
59 Lawrence Wright, *Twins: Genes, Environment and the Mystery of Identity* (London: Weidenfeld & Nicolson, 1997), 78.
60 Samantha Frost, 'Ten Theses on the Subject of Biology and Politics: Conceptual, Methodological, and Biopolitical Considerations', in *The Palgrave Handbook of Biology and Society*, ed. Maurizio Meloni, John Cromby, Des Fitzgerald and Stephanie Lloyd (New York: Palgrave, 2018), 902.
61 Frost, 'Ten Theses on the Subject of Biology and Politics', 902.
62 Timothy Clark, 'Scale', in *Telemorphosis: Theory in the Era of Climate Change, Vol. 1*, ed. Ton Cohen (Ann Arbor: Open Humanities Press), 150.
63 Stacy Alaimo, *Exposed: Environmental Politics & Pleasure in Posthuman Times* (Minneapolis: University of Minnesota Press, 2016), 11.
64 Anna Lowenhaupt Tsing, *The Mushroom at the End of the World: On the Possibility of Life in Capitalist Ruins* (Princeton and Oxford: Princeton University Press, 2015), 140.
65 Tsing, *The Mushroom at the End of the World*, 140.
66 Tsing, *The Mushroom at the End of the World*, 38.
67 Mark Fisher, *The Weird and the Eerie* (London: Repeater Books, 2016), 12.
68 Fisher, *The Weird and the Eerie*, 12.
69 Douglas Hofstadter, *I am a Strange Loop* (New York: BasicBooks, 2007).
70 Timothy Morton, *Dark Ecology: For a Logic of Future Coexistence* (New York: Columbia University Press, 2016), 25.
71 Morton, *Dark Ecology*, 16
72 Morton, *Dark Ecology*, 16
73 Fisher, *The Weird and the Eerie*, 43.
74 The narrative device most commonly used to evoke strange temporal loops is time travel. Time travel narratives often employ a twist on the gothic double motif exemplified by Poe's 'William Wilson'. If a time traveller meets a future or past version of her or himself it usually ends in the death of one of them, just as in Poe's story the protagonist's double is a harbinger of his death. In Terry Gilliam's *12 Monkeys* (1995) as well as its inspiration, Chris Marker's short film *La Jetée* (1962), the protagonist is haunted by the memory of seeing a man being shot in an airport. The ending reveals that the man was a future version of himself as if the excessive nature of this temporal paradox can end only in death. The 2012 thriller

Looper, directed by Rian Johnson, fuses the tropes of time travel and the gothic double in an even more explicit way. The title refers to a term used to describe assassins in 2074 who use time travel as a way to safely dispose of the enemies of their criminal employers. The major catch in this lucrative line of work is that, in order to safeguard the crime syndicate's dirty secrets, the assassin's final victim must be an older version of himself. Loopers must close the temporal loop with their own death.

75 Robert Plomin, *Blueprint: How DNA Makes Us Who We Are* (London: Penguin Books, 2018).
76 Shelley Jackson, *Half Life: A Novel* (New York: HarperCollins e-books, 2009), 6.
77 Jackson, *Half Life*, 47.
78 Jackson, *Half Life*, 436.
79 Amanda Greer, 'Absence, Play, and the Antidetective Story: Shelley Jackson's *Half Life*', *Critique: Studies in Contemporary Fiction* 59:2 (2017): 172.
80 Jackson, *Half Life*, 437.
81 Jackson, *Half Life*, 433.
82 Stéphane Vanderhaeghe describes the novel as having a 'recursive pattern.' 'How to Unread Shelley Jackson', *Transatlantica* 2 (2010): 4.
83 Greer, 'Absence, Play, and the Antidetective Story', 177. Greer is borrowing the term 'flickering' from Brian McHale's account of the trope of 'worlds under erasure' in his book *Postmodernist Fiction*. Here, McHale is referring to Jacques Derrida's strategy of placing words 'sous rature,' physically crossed out and yet still visible in a typographical performance of the way key terms in Western metaphysics (such as presence and objecthood) continue to prop up philosophical discourse despite being illegitimate. Postmodernist authors use a similar strategy when they bring fictional worlds into being while simultaneously erasing those fictional worlds. In a way that is particularly pertinent to *Half Life*, McHale argues that '[t]his violation of the law of the excluded middle becomes especially crucial when it occurs at one particularly sensitive point in the text, namely its ending.' He describes the 'ontic spheres' created by these strategies of 'self-erasure' as 'flickering worlds.' Brian McHale, *Postmodernist Fiction* (London: Routledge, 1987), 101.
84 Nicole Seymour, *Strange Natures: Futurity, Empathy, and the Queer Ecological Imagination* (Champaign: Illinois Scholarship Online, 2017), 153.
85 Jackson, *Half Life*, 437.
86 Eugene Thacker, 'Dark Media', in *Excommunication: Three Inquiries in Media and Mediation*, ed. Alexander R. Galloway, Eugene Thacker and McKenzie Wark (Chicago and London: The University of Chicago Press, 2014), 113.
87 Thacker, 'Dark Media', 111.
88 David Punday, 'Involvement, Interruption, and Inevitability: Melancholy as an Aesthetic Principle in Game Narratives', *SubStance* 105:33–3 (2004): 88.

89 Punday, 'Involvement, Interruption, and Inevitability', 99.
90 N. Katherine Hayles, 'Flickering Connectivities in Shelley Jackson's *Patchwork Girl*: The Importance of Media-Specific Analysis', *Postmodern Culture* 10:2 (2000): 13.
91 Hayles, 'Flickering Connectivities in Shelley Jackson's *Patchwork Girl*', 17.
92 Hui, 'Machine and Ecology', 57.
93 Hui, 'Machine and Ecology', 58.
94 Florence Chiew and Alison Barnwell, 'Methodological Intimacies and the Figure of the Twins,' *The Sociological Review Monographs* 67:2 (2019): 473.
95 Danowski and Viveiros de Castro, *The Ends of the World*, 14.
96 Danowski and Viveiros de Castro, *The Ends of the World*, 14.

4

Twinning in black futurism

The photograph 'Aljana Moons Twins Carriage', by artist Alexis Peskine, is a portrait of twin brothers taken by the ocean in Senegal. [See Figure 4.1] Both boys are sitting on a wooden platform, which we presume belongs to the carriage of the title. One faces the camera and looks straight into the lens, while the other sits in profile. They are dressed identically in suits made out of repurposed rice sacks with shiny tin shoulder pads, belts and helmets fashioned out of pureed tomato cans with the branding (Dieg Bou Diar) clearly visible. The shape of the helmets and the severity of their demeanours give the impression of two astronauts setting off on a space voyage, though this impression is belied by the artisanal construction of their outfits and the low-tech location. This image of youth making a claim on the future despite their circumstances exhibits one of the central critical strategies associated with Afrofuturism.[1] The photograph places conflicting temporal markers in opposition with each other: the futurism of the space-aged suits and the tradition embodied by a mode of transport (the horse-drawn carriage) that has been used for centuries; the time of techno-utopian progress central to the imaginary of space travel and the cyclical time of recycling and tidal patterns.

Twinship, as it is deployed in this image, embodies these tensions. On the one hand, the twins evoke modern modes of capitalist production. Clad identically in the repeated Dieg Bou Diar logos, the brothers appear as, in Hillel Schwartz's words, a 'synechdoche of mass production', an extension of the logic of serialized replication into human life itself.[2] On the other hand, the photograph draws on more culturally specific associations with the role of twins in West African spirituality. Across the region, twins have been both worshipped and demonized; viewed as disruptions of the natural and social orders that require containment by ritual. The association between the twins and ritualistic traditions is reinforced in Peskine's short film 'Aljana Moon' which shifts focus between the futuristic aesthetic of space travel and documentary footage of coming-of-age rituals.

Figure 4.1 'Aljana Moons III' by Alexis Peskine, 2015. Reproduced with the permission of October Gallery.

Framed by both Afrofuturism and spiritual ritual, twinship in the photograph draws attention to how the oppositions in the image, in Robert Stam's words, are 'less a question of juxtaposing the archaic and the modern than deploying the archaic in order, paradoxically, to modernize, in a dissonant temporality which combines a past imaginary communitas with an equally imaginary future utopia'.[3] The twins are a tool of what Kodwo Eshun terms the 'chronopolitical act' of Afrofuturism.[4]

'Aljana Moons Twins Carriage' is illustrative of how twins have been used to expose and challenge dominant associations between blackness and technology. In the short film, a sequence in which the twins walk along the beach in their space suits is followed by footage of a carpenter planing wood wearing a t-shirt bearing the slogan 'Black Resistance 2.0'. As the only words in the film, they appear freighted with meaning and encourage the viewer to extrapolate on wider political implications of the conjunction between twinship and the Afrofuturist aesthetic, particularly in relation to digital network technologies of the social media era. We are encouraged to think of Afrofuturist twins as tools through which to rethink how blackness is performed, challenged and reaffirmed in the digital age. In this respect, the 'Aljana Moons' project belongs to influential scientific and artistic traditions that use twins to either reinforce or challenge the logic of racial thinking. We have seen how the eugenics movements in Europe and the Americas during the first decades of the twentieth century leaned heavily on twin studies to support their claims. All of the most influential popularizing works of eugenics theories contain sections on twins.

Karen Dillon identifies a tradition within black literature in the United States that draws on these associations between twins and eugenics to critique the role played by 'racial kinship boundaries' in the establishment of a national community.[5] Due to their strong associations with closeness and insularity, twins have long been used as metaphors for modes of community that forge similarity across difference. Dillon focuses on two novels published in the second half of the twentieth century – *dem* (1967) by William Melvin Kelley and Toni Morrison's 1998 novel *Paradise* – that, in contrast to this tradition, emphasize the differences within the similarities of twin relationships to 'problematize the significance of racial legibility and allegiance as indications of community membership'.[6] Both novels, according to Dillon, narrate a transition from the Civil Rights era desire for integration with white society to an emphasis on opposition that was most fully embodied by the black Power movement. However, they do so in very different ways. In *dem*, a case of superfecundation, which occurs when a woman gives birth to twins with different biological fathers, leaves a middle-class white family with one white child and one black child. This narrative device functions as a satirical thought experiment that isolates the agency of race from other factors in the production of social inequalities. In this respect, the novel echoes Mark Twain's use of twins in *Pudd'nhead Wilson* to, in Dillon's words, 'investigate discrepancies between the law and social custom regarding definitions of [racial] identity'.[7] Kelley uses this strategy to present racial integration as impossible and undesirable. In Morrison's *Paradise*, by contrast, a twin relationship is used as a metaphor for race-based community. The fact that the stifling insularity of this relationship ends in stasis and entropic decay 'reveals the oppressive nature of similarity in restrictive conceptions of black identity that dictate communal belonging'.[8] Despite their differences, both novels draw on the metaphorical and allegorical potential of the twin bond in their interventions into racial politics.

The works of Afrofuturism and black speculative fiction explored in this chapter draw on the rich metaphorical and allegorical associations of twinship. However, in contrast to Kelley and Morrison, the Afrofuturist twins I examine evoke allegory only to undermine it. In Afrofuturism and black speculative fiction, I argue, twins are recruited into what Ian Baucom in *Specters of the Atlantic* has described as a 'counterallegorical practice' that exposes connections between allegory and the logic of speculative capitalism that was inaugurated with the transatlantic slave trade.[9] Afrofuturist twins operate as glitches within the allegorical systems that regulate blackness in the post-slavery context of the

African diaspora. In this, they echo a tendency within anthropological studies of twinship in West Africa to view twins as, in the words of Elisha Renne and Misty Bastian, 'occupy[ing] a position at the intersection of conflicting epistemologies and systems of power'.[10] The custom of twin infanticide practised by Igbo communities in Southern Nigeria up until the colonial period was a key point of conflict between local customs and the worldviews imposed by the British colonial system. As Bastian puts it, '[j]ust as multiple births (*umu ejime*) were considered by Igbo-speaking peoples an abomination (*nso ani*) against Ala/Ani, the earth deity, and liable for severe sanctions on both parents (especially mothers) and children, twin killing was an abomination against the Christian God in the eyes of the missions'.[11]

As an example of how the treatment of twins in West Africa reveals the 'often strained' relations between 'local understandings of what constitutes subjectivity, morality, and social value' and 'Western epistemologies and praxis', Renne and Bastian cite Chinua Achebe's *Things Fall Apart* (1958). In the novel, twins play a minor but crucial mediating role between the customs and beliefs of the Igbo tribal communities that are the focus of the narrative and the British missionary and colonial institutions that impose themselves on these communities. Rejection of the Igbo practice of twin infanticide works in the novel as both a central cause of the missionaries and key motivation among the tribal communities for turning their backs on their own customs. In a discussion with a tribal elder, one of the missionaries, speaking through an interpreter, evokes infanticide in his condemnation of the Igbo cosmology: 'All the gods you have named are not gods at all. They are gods of deceit who will tell you to kill your fellows and destroy innocent children.'[12] Similarly, disgust with twin infanticide is a key motivation among the early converts to Christianity. Among them are Nneka who 'had four previous pregnancies and childbirths. But each time she had borne twins, and they had been immediately thrown away'; and the protagonist Okonkwo's eldest son Nwoye whose 'young soul' is 'haunted' by 'the question of the twins crying in the bush'.[13] Twinship and the missionary church are viewed by the elders as equal threats to their kinship systems, entities that must be ejected to maintain its coherence. The fact that they both occupy the same liminal position in relation to Igbo kinship structures is made literally the case when the missionaries build their church in the 'Evil Forest', the 'bad bush' on the edges of the village where twins are taken to die. 'When one came to think about it, the Evil Forest was a fit home for such undesirable people [the church]. It was true they were rescuing twins from the bush, but they never brought them into the village. As far as the villagers were concerned, the twins still remained where they had been thrown away.'[14]

In his instructions to fellow artists on how to make a work of glitch art, Nick Briz tells his readers to 'take a familiar piece of technology & do something unfamiliar with it'.[15] For example, if you open a computer file with an application that was not intended for the file type, 'u're likely to get an error along the lines of: "sorry, unsupported file type"'.[16] However, with a bit of luck, the application might produce 'an unusual interpretation of the file'. In the process, this 'glitch' that emerges from the incompatibility between the file and the application draws attention to the rules governing both systems, the logics of exclusion that maintain their coherence, the interpretations through which they separate signal from noise. These glitches can also, in their unpredictability, open up glimpses of alternative systems. The texts and performances that I explore in this chapter all return to the incompatibility of twins staged in *Things Fall Apart* and reinterpret it as glitch. I start by discussing the forms of disjointed diaspora in *The Icarus Girl* before exploring the refusal of allegory performed by the twins in Tade Thompson's Wormwood Trilogy. I then look at the use of glitch aesthetics by dance duo Les Twins. The recursive processes of twinning are employed to loop together traumatic histories of slavery with speculative assemblages that articulate new connections between blackness and digital technologies.

Race as medium

Twins constitute a pervasive presence in Afrofuturist and black speculative fiction. As a figure that resonates strongly with both the technoscientific manipulation of human life (through the twins-qua-clones trope) and the porosity of boundaries between spiritual and material worlds in African cosmologies, they seem made-to-measure for a genre that sets out to explore the overlaps and intersections between these two discursive spheres. J. Griffith Rollefson points out that since its emergence, the concept of Afrofuturism has appealed to those interested in 'critiqu[ing] the reified distance between racialized fictions of *black magic* and *white science*'.[17] As important mediating figures in both of these discourses, twins provide a means of exposing both the commonalities (such as shared reliance on myth) and the glitchy ruptures of meaning that emerge from their contact. In Helen Oyeyemi's *The Icarus Girl* (2005), twinship constitutes the fulcrum between competing explanations of subjectivity. The narrative centres on eight-year-old Jessemy 'Jess' Harrison, the precocious child of an English father and Nigeria mother. The appearance of an imaginary friend in the form of trouble-making doppelgänger TillyTilly triggers a series of crises for both Jess and her family as she tries to forge an identity between the competing cultural influences of her mixed heritage.

The novel offers up two possible explanations for the appearance of TillyTilly to both Jess and the reader. The first is provided by a psychotherapist in London who looks to immediate family dynamics ('Jessamy, are you scared of your mum?') and encourages Jess to protect the boundaries of her unified subjectivity ('When your eyes are closed, you're inside yourself, and no one can get you there').[18] The second explanation derives from the Yoruba cosmology that Jess is exposed to when visiting her mother Sarah's family in Nigeria. Through TillyTilly, Jess learns that she was born a twin and her sister, Fern, died while she was being born. Against her father's wishes, Sarah did not take the usual ritual precautions prescribed by Yoruba spirituality when a twin dies. In Yoruba traditions the twin bond is believed to be so strong that the dead twin will attempt to drag the surviving twin back into the spirit world. The parents of the surviving twin must take all manner of precautions to stave off this threat. These include the construction of an anthropomorphic memorial statue of the dead twin, known as *ere ibeji*, to which the family must bestow the same attention and affection as the living twin. As Babatunde Lawal explains, the ere ibeji is thought to 'localize the soul of the deceased so as to maintain the spiritual bond between the living and the dead'.[19] If these measures are not taken then the dead twin will become what is known as *abiku*, a restless spirit caught between the living and the dead and will endlessly return to haunt the family and cause its ruin.

Through a narrative technique that has been described as 'Yoruba Gothic', the book is structured around the return of these spiritual beliefs that have been repressed from the family life the Harrisons have constructed for themselves in London.[20] Initially, twinship becomes a metaphor for the alienation and cultural hybridity of diasporic existence. Diana Adesola Mafe points out that *The Icarus Girl* echoes the connection made by Homi Bhabha between hybridity and the uncanny. In being split with her twin TillyTilly, Jess is, in Bhabha's words, 'less than one and double'.[21] In this respect, twinship in *The Icarus Girl* shares an allegorical role with a number of novels that explore the in-betweenness of existence in the African diaspora. In novels such as *The Second Life of Samuel Tyne* (2004) by Esi Edugyan and *26a* (2005) by Diana Evans, in the words of Brenda Cooper, 'the use of twins becomes a coded language for the writers' own splitting, doubling and questing for their identities in London or Alberta, as well as for their connection with Africa.'[22] However, in *The Icarus Girl*, while initially providing a channel of allegorical meaning, twinship ultimately constitutes an intractable blockage by introducing a rupture into systems of knowledge. Jess's troublesome twinship confounds the explanatory model, steeped in the language of ego psychology, provided by the psychotherapist in London. When he suggests that she should control her fear by maintaining the boundaries of her ego and retreating to an 'inner space' ('You can just *be* scared and then stop. Nothing happens in between') Jess cannot square this account of subjectivity with her experience of being a twin that has been framed by Yoruba beliefs. 'But what about a twin, a twin who knew everything because she was another you? Could she do something in that time in between?'[23] The fact that twinship occupies a point of intersection between conflicting epistemologies is emphasized in the last sequence in the novel in which Jess is recuperating in a hospital in Nigeria after a car crash and her grandfather, with the encouragement of the doctors, places an ere ibeji statue in the corner of the room.

The Icarus Girl presents this intersection between 'black magic' and 'white science' as a glitch space. Shocked at Jess's inexplicable knowledge that she had a twin who died in childbirth, Sarah exclaims: 'Three worlds! Jess lives in three worlds. She lives in this world, and she lives in the spirit world, and she lives in the Bush. She's *abiku*, she always would have known!'[24] It is this 'third space', the Bush, that TillyTilly occupies and where she repeatedly attempts to lure Jess. When she first appears during a visit to Sarah's father's house in Nigeria, TillyTilly wakes Jess in the middle of the night and takes her to an amusement park. The two girls break in and have the place to themselves, playing into the night before TillyTilly abruptly disappears, leaving Jess to negotiate the long walk home alone. The park is a key variation on the liminal Bush space to which twins are banished in *Things Fall Apart*.[25] The park is described in terms of uncanny animation. The machines and playground systems apparently operate autonomously. 'It was as if the amusement park was *alive*.'[26] The inorganic liveliness of the park and the way it confounds distinctions between autonomy and animation reworks Yoruba Bush myth for the technocultural present of the novel. In her reading of *The Icarus Girl* as a work of 'postcolonial female gothic', Diana Adesola Mafe argues that the amusement park Bush space provides the stage for Oyeyemi's vision of a 'futuristic postcolonial diaspora', placing the novel within an Afrofuturist tradition.[27] Within the narrative of the novel, it foreshadows the glitch space that TillyTilly occupies once Jess and her family have returned to London. These encounters are described in the language of glitchy mediation. When TillyTilly tries to explain what it is like to be in the Bush, the interface between the two girls seems to breakdown, the audio slipping out of sync with the imagery. 'Jess watched as Tilly's lips moved noiselessly for a few moments, as if she has forgotten how to speak, or what the words were.'[28] When Jess is drawn into the Bush, she experiences it as 'that gap of perception between what is really happening to a person and what others think is happening'.[29] At the height of her crisis, Jess speaks on the telephone with her grandfather and experiences a desynchronization as she 'grimace[s] at the echo on the line'.[30]

In *Seeing through Race*, W.J.T. Mitchell makes a connection between the decline of racial identity in the 1990s (the idea that Western society was moving into a 'post-racial' phase) and the erasure of medium specificity by the rise of digital culture. For Mitchell, race is itself a medium. It is not something that we '*look at*' but rather something that we '*see through*' – an 'intervening substance' that 'mediates,' and therefore produces, 'socio-political relations'.[31] Since conceptualizing race as a medium avoids the twin pitfalls of biological essentialism and dismissing it as an ideological fantasy, it is a useful lens through which to study the persistence of racial thinking. It also provides a way of exploring how the concept of being 'post-racial' is intertwined with the discourse about being 'post-medium'. The idea that digital technology obliterates the specificities of pre-existing forms of mediation (including print, audio and visual media) by abstracting them to a series of 1s and 0s found its corollary in the notion that Western societies have evolved beyond the need for racial distinctions. The return to medium specificity in a range of cultural practices – from the resurgence of vinyl, cassettes and independent printing to the rise of academic interest in the material conditions of mediation in academic methodologies such as book history and media archaeology – must be seen in conjunction with a persistence in viewing the agency of race not as a biological determinant but, to use Mitchell's definition of the concept of medium, as a 'material social practice'.[32] Through its construction of twinship as a form of glitchy, weird mediation, the Afrofuturist writers and performers discussed in this chapter, I argue, emphasize the agency of both race and mediation and, furthermore, present the two as being intricately intertwined.

The appearance of the doppelgänger twin TillyTilly in *The Icarus Girl* triggers a heightened awareness in Jess of the role played by race in her construction of social relationships. Commenting on one of Jess's increasingly frequent breakdowns, a fellow student at her school comments: 'Maybe Jessamy has all these "attacks" because she can't make up her mind whether she's black or white!'[33] Whereas Oyeyemi presents Jess's twin as the embodiment of her growing awareness of race, other writers have presented twinship as a form of otherness that doubles and further emphasizes the otherness of race. In Dillon's analysis of Morrison's novel *Paradise*, for example, she focuses on a scene in which black twin brothers catch the eyes of some white townsfolk. '[S]ome whitemen, amused by the double faces, encouraged the brothers to dance.' When one brother, Tea, complies with the demand and the other brother, Coffee, refuses and is shot in the foot for his resistance, their diverging reactions to the attack irrevocably drive a wedge between them. For Dillon, '[t]his scene simultaneously illustrates the white gaze on racial otherness and a cultural gaze on the otherness of twins'.[34] The brother who refused to dance 'feels the fact of his blackness *and* his twinship in a double gaze: he watches the white men gaze upon Tea and experiences the gaze vicariously through his twin'.[35] In Marjorie Wallace's biographical account of the 'silent twins' June and Jennifer Gibbons, the sisters' status as 'identical' twins emphasizes their racial difference from the predominantly white inhabitants of their neighbourhood in rural Wales. Daughters of Caribbean immigrants, June and Jennifer were born in Barbados and grew up in rural Wales during the 1980s. When, following a series of failed encounters with education institutions, the girls started committing a series of petty crimes they were sent to Broadmoor hospital where they spent 11 years. In Wallace's account in *The Silent Twins*, the otherness of twinship seems to legitimize the use of racially loaded terms of description as the sisters are described as 'zombies' or compared to 'voodoo dolls'.[36] In contrast to the 'silent twins', in *The Icarus Girl* twinship is employed to insist on race as a form of mediation that can be manipulated in an attitude that is shared by much Afrofuturist work. If the 'idol of race cannot be smashed', Mitchell argues, it is better to adhere to Nietzsche's wise advice and '"sound the idol with a hammer, as with a tuning fork" to listen to its hollow resonances and retune them in accordance with a music to come'.[37]

Race and species being

In Tade Thompson's Wormwood books, twins are used to 'retune' the medium of race for a network age. The trilogy is a more explicit and knowing intervention into the genre of Africanfuturist science fiction than *The Icarus Girl*. The first instalment, *Rosewater*, published in 2016, is set in rural Nigeria in 2066 where a vast and mysterious alien dome has emerged from the Earth. The dome communicates with the human inhabitants of Earth indirectly in two ways. Firstly, once a year an opening emerges and anybody ill or injured in the vicinity of the dome is healed. Often the dome produces jarring interpretations of 'healing' by resetting bones in the wrong way or reanimating corpses, and in the process producing a new species of zombie-like 'reanimates'. Secondly, the dome releases a vast net of fungal spores into the surrounding atmosphere that functions as a sophisticated communication system, allowing it to gather knowledge about its new habitat. A select number of humans are able to tap into this biotechnical information network, or 'xenosphere' as it becomes known, and use it as a 'telepathic' medium to infiltrate other people's thoughts and dreams. The first book of the trilogy focuses on one of these so-called 'sensitives' who works for a Nigerian intelligence service as he negotiates the systems of power that have configured around this alien entity. Through speculative and allegorical displacement, the trilogy interrogates the political implications of pervasive connectivity (the xenosphere becomes a highly efficient surveillance system) and probes the forms of subjectivity that result from these emerging modes of connection. The novels proliferate with symbiotic entities and examples of distributed cognition, from the 'sensitives' who can manipulate peoples' thoughts to the method of alien 'invasion' itself which aims to bond with humans rather than replace them.

Twins seem to multiply through the pages of the Wormwood trilogy. In an ironic echo of the role of Romulus and Remus in the mythic founding of Rome, two of the major districts of the city of Rosewater that has sprung up around the dome are named after twin brothers Taiwo and Kehinde, gangsters who were hired by the first political leaders to control the growing populace. But it is in the second book, *The Rosewater Insurrection* (2019), when twinning becomes a key metaphor for both symbiotic life and mediation. It is in this book that we learn the history of the alien invasion. The xenoforms (the fungal spores that make up the xenosphere) were shot into space as emissaries by the surviving inhabitants of a dying alien species, whose earth has become uninhabitably polluted, in search of a potential new home planet. They were carried as parasites by a host species known as 'footholders', described as 'organic blobs' that need to bond with another species to acquire sentience. The parasitic xenoforms are 'twinned' with other 'sender-receiver' xenoforms that are housed on a moon of the home planet. Information is sent between them through the 'Spooky Action at a Distance' of quantum entanglement.[38] Facing total extinction, the 'Homians', as the alien species are known, have made 'imprints' of themselves by uploading their 'memories and consciousness' onto 'bio-mechanical storage units'.[39] If the footholders (and their xenoform riders) find a habitable planet, they will send a message to the home planet triggering a mechanism that will send the Homian consciousnesses to be reawakened in their new home. The narrative of *Insurrection* focuses on the impact of a further organic entity, strain-516, that was sent with the 'footholders' to check their growth and make sure they do not entirely overrun the potential host planet. This plan goes awry when strain-516 bonds with a particularly malevolent human and the resulting symbiotic entity threatens to obliterate the dome and sabotage the Homian colonizing mission.

The 'twinning' of the xenoforms that initiates this interplanetary colonization becomes a model for the other forms of symbiotic being in the novel. One of the protagonists of *Insurrection* is a footholder called Anthony who, using information gathered through the xenosphere, has cloned the form of a human to become an intermediary between the Homians and the inhabitants of the potential host planet, Earth. When Anthony leaves the dome in search of strain-516, he encounters one of several giant ere ibeji statues that were commissioned by Taiwo and Kehinde. Anthony learns from a local about the 'quasi-divine status of twins to the Yoruba' and that the first twins in Yoruba legend were abiku, 'unborn spirits [who] keep coming back in an endless cycle'.[40] The ere ibeji carvings, he learns, are viewed as a 'repository of the spirit of the twin'.[41] The figure of ere ibeji is immediately legible within the narrative as a metaphor for subjective entanglement. Just as the existence of the surviving twin is entangled with the spirit of the dead twin, so the existence of all the Homian envoys is enmeshed with other organisms. Directly after his encounter with the ere ibeji statue, Anthony locates strain-516 where it first bonded with its host human. What he finds there is a monstrous mirror image of his own entwinement. 'The vines twist without warning, corralling the stems, parting the green sea until the body at the centre is exposed. It is emaciated, cadaverous, mottled light and dark green, punctured by roots and vines on most of its surface area. The milky eyes are open and it frowns.'[42] Furthermore, if twinship is set up as a metaphor for entwinement, it is a form of entwinement that is not premised on binaries but rather follows a fractal logic. The Anthony in *Insurrection* is the last in a series of clones that endlessly reproduce themselves. When he awakens, he finds himself 'surrounded by abandoned versions of himself'.[43] The proliferation of the clones foreshadows the growth emanating from the strain-516-human hybrid.

As well as entwinement, the account of ere ibeji delivered to Anthony also provides a metaphor for biotechnical mediation. The novel ends with some of the human inhabitants of Rosewater helping Anthony and his fellow footholders to eradicate the threat of strain-516 and help them transmit the Homians to Earth. One plan is to use the zombie 'reanimates' as receptacles for the Homian consciousnesses. Explaining the idea, one character compares this possible solution to the containment of twin spirits within ere ibeji statues. The reanimates, she explains, 'are empty of souls, like the wooden carving'.[44] However, in a way that echoes *The Icarus Girl*, twinship as a form of mediation in *Insurrection* is characterized by its glitchiness. Responding to the idea of using the reanimates as vessels for the Homian souls, the mayor points out that the solution fails to take into account the role of embodiment in the 'illusion' of consciousness. 'What you've encoded is memory, and personhood is not just memories. Personhood is embodied.'[45] If the role of ere ibeji in Yoruba myth provides a model of transmission of 'spirit' between human bodies, it is a form of transmission that is far from noise-free. The founding twins themselves, Kaiwo and Kehinde, use their twinship to block the flow of information in the xenosphere. All inhabitants of Rosewater are implanted with an ID chip that connects to the xenosphere, allowing the state to monitor closely the actions and thoughts of its citizens. The twins both implant themselves with an identical second bootleg implant that serves to both easily transmit information between the brothers and 'jam the legitimate [implant] during the commission of a crime, leading to a non-person'.[46]

The racial implications of the ere ibeji media metaphor are clear. One of the many science fiction tropes that Thompson satirizes is the use of alien invasion narratives as allegories for colonialism. In this respect, the Wormwood trilogy echoes the South African film *District 9* (2009) in which alien visitors are subjected to systemic racist abuse echoing that of apartheid. The otherness of species thinking is displaced onto the otherness of race. Generally, the twist on alien-invasion-as-colonialism trope carried out by Thompson is that the process of colonization described in the trilogy is not one of violent imposition but rather intimate entwinement. As Steven Shaviro puts it in his review of *Rosewater*, '[t]he coming disaster is, and will continue to be, an *intimate* one. It will be something for which "we will all be present," even as we are devastated by it.'[47] Binary colonial discourses of othering are replaced by fine gradations along the scale of animacy. The use of the ere ibeji statue as a metaphor for using the 'reanimates' as 'receptacles' for Homian life evokes what Zakiyyah Iman Jackson describes as the history within Enlightenment discourse of 'blackness's bestialization and thingification: the process of imagining black people as an empty vessel, a nonbeing, a nothing, an ontological zero'.[48] Like blackness in the scientific and philosophical traditions explored by Jackson, in *Insurrection* the 'reanimates' constitute a kind of limit-case of subjectivity. In *Rosewater*, they are deemed beyond the pale by the state and are systematically eradicated. In *Insurrection*, following a spate of hate crimes, this status is being contested by a movement that seeks to extend citizenship rights to 'alternatively animated individuals', a term they prefer over the 'reanimates' label.[49] Their objective is to achieve, on behalf of the 'reanimates', inclusion into the dominant conceptions of personhood. The movement is heavily satirized in the novel, the humour deriving from displacing the language of inclusion and recognition of identity politics onto relations between species.

This satirical approach throws into relief the treatment of race in the novel itself. The footholder Anthony treats race as a medium that can be manipulated. 'The colour thing confuses him so. It's a human thing. Near identical DNA, yet they discriminate against each other based on the divisions of white light and the degree of protrusion of the jaw or the shape of the eyes or nose.'[50] Although he does not fully understand it, he attempts to use perceptions of racial differences to mediate his presence to others. When no cars stop for him while hitchhiking, for instance, he thinks that making his skin darker (to resemble more closely a human friend he deems trustworthy) will improve his chances. Even amidst the vertiginous proliferation of species driven by the Homian invasion, race is a pervasive 'intervening substance' that mediates social existence and determines differential access to the category of personhood. Crucially, race is presented as intertwined with differences between species and the different levels of animation constituted by different agents within the various symbiotic partnerships. Unlike the pro-reanimate activists, *Insurrection* is not seeking for its characters inclusion into the definitions of the human that were produced by racist colonial discourse. Rather, it presents blackness alongside a vast and fractally proliferating range of species difference to challenge these categories. In this respect, the critique carried out by the Wormwood books coincides with the works of African diasporic culture addressed by Jackson. Whereas such works have often been interpreted as 'a plea for human recognition' in 'reaction to racialization', works by writers such as Nalo Hopkinson confront the history of the 'bestialization and thingification' of blackness not in the name of inclusion within dominant categories of the human but to displace them. These works, 'creatively respond to the animalization of black(ened) being by generating a critical praxis of being, paradigms of relationality, and epistemologies that alternately expose, alter, or reject [...] the racialization of the human–animal distinction found in Western science and philosophy'.[51]

The use of twins in *Insurrection* is central to the articulation of this discourse. As we have seen, the ere ibeji statue (seen as a 'receptacle of the souls of dead twins') is used as a metaphor for the enslaved black body viewed as an empty vessel, life reduced to commodity in, as Baucom pointed out, a social enactment of the logic of allegory. However, this allegorical use of twinship is repeatedly distorted and undermined by the associations between twins and inter-species entanglement. The incompatibility between these two allegorical uses of twinship in the novel functions as a glitch. Twins sit at the intersection between these two allegorical regimes and in each they function as either the key to the system or a point of failure and excess of meaning depending from which perspective they are viewed. This use of twins is in keeping with the violence the Wormwood trilogy inflicts on allegorical interpretations more generally. The Homian invasion evokes a number of historical analogues (neoliberalism, European colonialism, etc.) without any of them fitting perfectly. Thompson's twins, then, are key tools in his 'counter-allegorical practice'.

Glitch performance in Les Twins

Dillon points out that the cultural history of twins, as objects of scientific scrutiny or marvel, is 'inextricably linked with spectacle, experimentation, and eugenics'.[52] The twinship of the protagonists of *Paradise* redoubles the otherness of their blackness, subjecting them to a double gaze. Similarly, as we have seen, in Wallace's *The Silent Twins*, the fact that the sisters are identical twins seems to legitimize their status as objects to be looked at. However, there is a countertendency of performers who use the double otherness of twinship and blackness to critically frame both as forms of mediation. Among the best-known twin performers are Millie and Christine McCoy, conjoined black twins born into slavery in Carolina in 1851. The sisters, whose act consisted of singing (they were sometimes referred to as the 'two-headed nightingale') and conversing with the audience, had a successful career that included tours to Europe and a stint with P.T. Barnum's circus where they built on the intrigue generated by their more famous conjoined predecessors Chang and Eng Bunker. Their early experiences of touring in the 1850s exemplify the point of intersection between scientific scrutiny and popular spectacle that Millie and Christine's managers sought to target. As Joanne Martell explains in her biography of the twins, when they arrived in a new town, '[b]efore opening the doors to the general public, they invited local medical men to come satisfy their scientific curiosity'.[53] This would serve the dual function of authenticating the sisters as being genuine conjoined twins and drumming up curiosity to boost ticket sales. Their value to science would be used to stoke public curiosity (an early newspaper headline claimed that the sisters 'have been pronounced by Physicians the most interesting specimen of Humanity ever seen'), while the interest demonstrated by scientists was clearly influenced by their popular renown.[54]

The language of race was also exploited by the managers of Millie and Christine to promote their shows. When they were exhibited in P.T. Barnum's

American Museum in New York in 1854 at the age of three, they were billed as 'the Celebrated African United Twins'.[55] The description as African worked in two ways by, on the one hand, emphasizing their exoticism and, on the other, evoking the fact that they were slaves to legitimize their status as spectacle. At this level, race was evoked to justify their existence as objects of exchange, a reality that was repeatedly emphasized during their lives by the fact that they were stolen as children and dragged through various legal processes to establish their rightful owners. The language of race was also used to position them carefully in relation to dominant conceptions of the human. As the early newspaper headlines attest, the humanity of the twins was a key component of the performance. The main role played by the doctors and the anthropometrics they deployed was to confirm the fact that they were human, indeed the 'most interesting specimen of Humanity even seen'. But if they were presented as being human, it was a particularly bestialized form of humanity. In the advert for their appearance in Barnum's American Museum, an image of the 'African United Twin' occupies the same column with the 'Only Living Rhinoceros' and the 'Prince of Serpents', a 28-foot-long boa constrictor. The conjoined status of the twins reveals the fact that, as Jackson explains, '[d]iscourses on nonhuman animals and animalized humans are forged through each other; they reflect and refract each other for the purposes of producing an idealized and teleological conception of "the human"'.[56]

However, although the publicity campaigns run by managers and theatre-owners continued to describe the twins as 'African', it is striking how little race is mentioned in journalistic and popular accounts of their shows. When their blackness is mentioned it is more often than not downplayed by describing the girls as having a 'fair Creole complexion'.[57] This reception is testament to the fact that the twins occupied the intersection of different discourses. Their supposed value to science, coupled with the intelligence they displayed in conversation (a quality almost universally commented upon), seems to have whitened them in the eyes of their spectators. As a result, the twins sat awkwardly on the spectrum of animation that undergirds dominant conceptions of humanity. As enslaved beings and objects of exchange among their white managers and audience members, they were positioned as inanimate or at least animated to the same degree as the animals alongside which they were exhibited as toddlers in Barnum's museum. The main effect of the performances themselves, however, was to display their high level of animacy. Not only did the sisters have a 'fair Creole complexion' but they also possessed 'an animation that is truly attractive'.[58] From the accounts collected by Martell, the degree of animation ascribed to the sisters seems to differ from what Sianne Ngai terms 'animatedness'. In *Ugly Feelings*, Ngai identifies a cultural discourse rooted in the period of slavery that 'imagine[d] the racialized subject as an excessively emotional and expressive subject'.[59] This racialized figure of the black subject becomes the puppet for externally imposed emotions that she or he is not fully in control of has fed into a range of black performance genres. By contrast, Millie and Christine are presented as in control of their emotions, self-animated and self-possessed in a way that clashes with their physical embodiment.

French hip-hop dancers and choreographers Laurent and Larry Bourgeois, who operate through the name 'Les Twins', also use their twinship to intervene into dominant associations between race and animation. The brothers rose to international fame through a mix of reality TV talent competitions and viral YouTube videos, the most significant of which being an 8-minute-long clip of a dance tour uploaded in 2010. The video exhibits the brothers' characteristic blend of hip-hop dance styles and genres with fragmentary narrative, punctuated by moments of synchronicity when the twins mirror each other's movements with eerie precision. Their use of popping and locking techniques occupies an ambiguous space between autonomy and automation, as if the brothers are alternately resisting and surrendering themselves to an external agency that controls their movements. The abrupt shifts between rhythm and staccato counter-rhythm as well as between story-driven choreography and asignifying movement evoke an aesthetics of glitch. The first section of the San Diego dance, for instance, plays out against the soundtrack of a remixed version of Michael Jackson's 'Whatever Happens' (2001) as the brothers are seemingly pulled in and out of a cinematic fight scene. In one sequence, Laurent throws Larry dramatically across the stage before getting caught in a glitchy looping movement looking around as if in horror of his own actions. In another, a stylized punch-up is broken up with alternating moments of fast and slow motion. The overall effect is of a glitching interface, a breakdown of the mediation that leaves the viewers flickering between channels.

As their collective name suggests, twinship is central to both the dancing style developed by Larry and Laurent as well as to their brand. Birgit Abels analyses Les Twins' performances as a case study of dance as a process that 'transcends the binaries of mind-body, inside-outside, form-content and immateriality', opening up a 'non-dichotomous realm' through which to 'render one's embeddedness in the world experienceable and to actualize one's becoming in this relational process'.[60] Dance, for Abels, is a form of 'dwelling' which, drawing on the ecological theories of Tim Ingold, she describes not as a static mode of being, but as 'an emergent and lived practice' that 'takes place' through a 'co-becoming' between and among 'humans and their environments'.[61] The way that Les Twins use their twinship in their performances makes them especially valuable as examples of dance as a relational 'dwelling'. Abels points out how the brothers 'regularly frame their responsiveness to one another in interviews in terms of their mutual connectivity, often stressing their bond as twins'.[62] It is the intuitive 'connectivity' that Laurent describes as 'telepathy' that Abels positions as a particularly acute form of 'choreographic thinking', a term that Erin Manning developed to refer to 'thought in movement', which is 'activated not solely in the body, but across the machinery itself – in the folds of the evolving architecture'.[63] In other words, for Abels, Larry and Laurent's performance of their twin bond as an acute form of connectivity exposes dance as a form of relational becoming.

Twinship is also key to the glitch aesthetic developed by Les Twins. Through their use of contrapuntal synchronicity they present themselves as drifting in and out of each other's control. The mirroring that takes place between them often follows a recursive pattern as one copies a style developed by the other who then riffs on and develops the copy. One of the main genres that they build on is that of the battle in which they mirror back their opponent's movements with improvements and play up to the contingencies and constraints of the immediate context, from unforeseen developments in the music to reactions or intrusions from spectators. There are countless videos filmed on smartphones of the brothers spontaneously reacting to unplanned circumstances in dance battles, either defending each other from aggressive spectators or mocking the abilities of their opponents. The spontaneity of this genre is reproduced in even their most choreographed of performances. Through a recursive pattern of mutual mirroring and development they create an impression of virtuosity, the sense that the viewer is watching a process rather than a finished product that is reproduceable. In other words, twinship is evoked not as a 'synechdoche of mass production' but as a model of cybernetic being.

It is because of their manipulation of glitch aesthetics that Laurent and Larry's careers have become associated with the rising popularity of Afrofuturism. The twins toured several times with Beyoncé during the 2010s, culminating in a cameo appearance in her headline act at the Coachella festival in 2018. The performance, the preparation of which is documented in the 2019 Netflix film *Homecoming*, draws heavily on Afrofuturist traditions that had been present in Beyoncé's work since her 2016 'visual album' *Lemonade*. The pyramid formation of the stage, for instance, echoes the reinvention of Egyptian mythology by 1970s funk musicians such as Sun Ra and Parliament Funkadelic. The jarring temporality of this future projection of ancient Egypt is matched by the costume design, which is best described as a space-aged variation on traditional college marching band outfits. At one point, a majorette takes centrestage wearing glittering silver hot pants and a baton that glows like a light sabre. When they first appear, Les Twins descend the centre of the pyramid dressed identically in shining silver hooded jackets out of which emerge silver laced tasselled tails as if part of a ceremonial garment. Their glitch dance reinforces the show's overall critical gesture of jamming the frequencies of white mass culture, while their synchronous movements repeat in microcosm Beyoncé's use of symmetry in her staging of Afrofuturist pageantry.

In the same year, Larry and Laurent were cast in *Men in Black: International*, spin-off from the successful 'Men in Black' franchise that was panned by the critics. The brothers' performance as alien villains repackages the glitch aesthetics that made them famous. Within the narrative, the twins play an alien entity that has enmeshed itself with human DNA and cloned the appearance of one of their victims. They have the ability to enter into the molecular structure of the material world and can cause the city streets to ripple and wave like the surface of water. At an intertextual level, their performance also echoes the twin villains of *The Matrix Reloaded* (2003) who also have the power to manipulate the informational fabric of reality. Their glitching existence at the intersection of different dimensions within the plot is reproduced at the level of the medium of genre cinema. However, the seeming doubleness of the twins in the film (emphasized by their identical outfits) does not function as a microcosm of the machinic repetition of tropes in genre narrative. Rather, it is enacting the looping structures of genre conventions, the complex mirroring and splitting that takes places as storylines and characters migrate across different media.

Gail A. McFarland argues that the use by Les Twins of the doppelgänger motif exposes the 'double consciousness' of commercial hip-hop dancing. McFarland draws on the work of various analyses of hip-hop dancing that make a connection with W.E.B. Dubois's 1903 theory of 'double consciousness', which he defines in a famous passage of *The Souls of Black Folk* as a 'sense of always looking at one's self through the eyes of others' and therefore '[o]ne ever feels his two-ness, an American, a Negro; two souls, two thoughts, two unreconciled strivings; two warring ideals in one dark body'.[64] For Elizabeth Alexander, for instance, commercial hip-hop dancing enacts a double consciousness by, on the one hand, functioning as a site of embodied collective memory of historical black experience and, on the other, performing an image of the black body 'which will be interpreted as exotic to the outside world'.[65] For McFarland, Les

Twins deploy the doppelgänger motif to emphasize the 'double consciousness' of hip-hop dancing. In some instances, this takes the form of each twin embodying a separate aspect of this double consciousness. In one of their most theatrical performances, staged for the 2017 televised talent competition World of Dance, one twin plays a wheelchair-bound homeless man begging for change while the other plays a man in a suit. A fight ensues when the businessman refuses to acknowledge the beggar. Considering both the rare use of costume and the intense commercialization of the performance's context, the altercation is clearly presented as an allegory for the struggle over the dance form's soul.

However, the trope of the double is rarely evoked in such a clear allegorical manner. Rather, Les Twins more often alternate between doubling each other and presenting each individual performer as being internally doubled or split. The frames of reference for their movements shift abruptly as the dancers seem to move in and out of different spatio-temporal contexts. This is most evident in the use of violence by Larry and Laurent. Les Twins performances often contain sequences of highly stylized fighting as if the performers have been momentarily possessed by the choreographed actions of a kung fu movie being played through a glitching interface. Although the antagonism most often takes place between the brothers, at moments it becomes internalized by each dancer as they appear to strike themselves. In another of several performances they put on for the 2017 World of Dance competition, the twins dance to 'Never Know' by the rapper 6LACK from Atlanta, Georgia. Their movements veer between interpreting the lyrics in quite a literal way and what they describe as 'showing' the music in a more abstract way. One key lyric narrates how 6LACK's music career has diverted his life from a path of criminality: 'Ran through these streets, ran through these beats. Ran through these thoughts, ran through these hearts. I was in the dark till I caught the spark.'[66] Over the words 'ran through these beats', which is repeated twice in the song, both brothers appear to hit each other in the face in an extension of the trope of self-inflicted violence that features in a number of their performances for World of Dance. The action enacts the double meaning of the word 'beats', which refers both to musical beats and the act of hitting or being hit. But Larry and Laurent's interpretation also expands the range of possible meanings applying to the word in the context. At the moment of imaginary impact, the 6LACK track seems to become momentarily stuck as if the physical act of violence brings with it a more symbolic violence that affects the music itself. It is also presenting their dance as an act of interpretation that violently intervenes into the music rather than passively consuming it.

However, in light of Alexander's reading of hip-hop dancing as an act of embodied witnessing of historical black experience, the act of self-inflicted violence also places the ambiguities of 6LACK's public persona within a longer history of double consciousness. By becoming a commercial success, 6LACK is feeding a system that fetishes black bodies as spectacle while reproducing the systemic inequalities that he was trying to transcend. So, by rendering his own body a commodity, 6LACK is repeating the violence he suffered during his urban upbringing by submitting himself to the violence of commodification. However, by drawing attention to this ambiguity, Les Twins are not criticizing 6LACK. Rather, they are staging an intervention into the history of black performance that runs back to slavery. In his book *In the Breaks*, Fred Moten explores the ways through which the history of slavery manifests itself as the founding trauma of radical black performance traditions of the twentieth century. The starting point of black performance, he argues, is the 'historical reality of commodities who spoke – of laborers who were commodities before, as it were, the abstraction of labor power from their bodies and who continue to pass on this material heritage across the divide that separates slavery and "freedom".[67] The improvisatory jazz performances of the likes of John Coltrane and Ornette Coleman carry out a disruption of the division between (white, universal) subject and (black, bestial) object that undergirded and legitimized the production of human commodities through the system of slavery. This takes place through an insistence on the material conditions of mediation that displaces 'the oppositions of speech and writing, and spirit and matter'.[68] The insistence on race in these performances, Moten argues, is 'not in the interest of a nostalgic suturing of wounded kinship'.[69] Rather, it is evoked to explore how 'this irrepressibly inscriptive, reproductive, and resistant material objecthood' (or 'animateriality') might be deployed to question and reframe the 'exclusionary brotherhoods' of black performance.[70]

Moten's concept of the disruptive materiality of black performance provides a useful framework for examining Larry and Laurent's deployment of glitch aesthetics in their intervention into the discourse of Afrofuturism. The twins present themselves as glitch doppelgängers emerging from the intersection of the violent legacies of slavery and the networked conditions of digital spectacle. Their performances are poised between evoking this traumatic history (evoking the theory of double consciousness through the trope of self-inflicted violence) and anticipating the mediation and reception of the spectacle they produce. For twinship is once again presented as a form of weird media. Abels draws on Mark Butler's account of Electronic Dance Music performance in *Unlocking the Groove* to describe Larry and Laurent as occupying a point of mediation between the producers and consumers of the music. Through their interpretative act of 'showing' the music, they are positioning themselves as the 'performing audience'.[71] However, their performances do more than just help the audience interpret the music or hype the crowd up. Rather, they present the live performance as just one node of a network of digitally networked mediation. Their eclectic approach to both style and genre means that their performances resemble the dance equivalents of what Hiroki Azuma describes as 'database narratives'.[72] Their vast array of references assumes knowledge on the part of the viewer of an unruly network of dance gifs and movie memes. As a 'performing audience' of the music, the work of assemblage and interpretation carried out by Les Twins anticipates the work of very active consumption (or 'prosumption') that will be carried out by fans once the videos have been uploaded to YouTube through the editing and reuploading of clips, identification of soundtracks and production of memes. Their use of popping and locking techniques to evoke glitch aesthetics presents their performances as already mediated – already the product of endless repetition and sharing, as if the viral spread of each video is already contained within the choreography itself. The recursive looping of each other's movements within the dance anticipates and conditions the recursive development of their brand between audience reception, media platforms and the music, fashion and film industries.

The staging of this recursive logic is another key aspect of their intervention into Afrofuturism. The twins' performances share their recursive approach with the techno music of Jeff Mills. tobias van Veen points out how the use of repeated sound patterns by Mills is connected to his embrace of Afrofuturist myth systems. His 'compositional focus on mechanical, industrial, and otherwise alien forms of sonic repetition' is a musical corollary of 'the Afrofuturist tradition of adopting alien, cyborg, or machinic identities that perform the double task of both allegorizing and transforming the embodied performance of blackness'.[73] The performance of process developed by the twins, as they slip in and out of rhythms, styles as well as narrative and comic personas, suggests what van Veen describes as 'a technology of the self that undertakes a transformative *becoming*'.[74] However, as we have seen, the Les Twins performances present this 'becoming' as been emergent from the intersection between the historical trauma of black experience and the speculative appropriation of digital technologies. Through their performance of recursion, the brothers carry out what critic Rollefson argues is the most valuable critical potential of Afrofuturism. By undermining the 'reified distance' between 'black magic' and 'white science', and between past traumas and speculative futures, Afrofuturism 'collapses the binary' between 'the ethnic nationalism of essentialist thought' and the 'free-floating identities of post-structuralism'.[75] The power of Les Twins' intervention is the expression of this Afrofuturist discourse through what Abels describes as the 'non-dichotomous realm' of dance.[76]

Conclusion: The weird transnationalism of Afrofuturist twinning

In *Afrofuturism 2.0*, Anderson and Jones argue that contemporary Afrofuturist narrative (and the discourse of what they call Astro-Blackness) traces the transition of a 'nation-state bound analog notion of blackness' to 'a digitized era toward and in tension with post-digital perspectives as a global response to the planetary and near planetary challenge facing black life'.[77] Afrofuturism 2.0 explores the 'technogenesis' of black identity in an area of transnational digital networks. The digital phase of 'black futurity', Anderson has argued elsewhere, is influenced by a range of factors that include 'the emergence of social media, the rising impact of climate change, populist movements, and the fraying of the global socio-political paradigm of American hegemony'.[78] The trope of Afrofuturist twinning has been used to explore the transnational technogenesis of blackness. It does so by pointing simultaneously 'backwards' to the configurations of colonialism while also serving as a lens through which to speculate about emerging transcorporeal configurations of information and media that function across the scales of the nation-state. If, as Nicholas Royle has pointed out, the motif of the uncanny emerged in late-Victorian gothic literature as a symptom of Empire, then weird aesthetics are symptomatic of the transnational hyper-connectivity of digital networks and planetary climate crisis.[79] With associations that look back to the uncanny through the trope of the doppelgänger, and forward to the weird through the logic of recursion, twins occupy a point of intersection between these two regimes of globalism.

As we have seen, a number of novelists have used twins to explore various aspects of the African diasporic experience. In *The Second Life of Samuel Tyne*, for instance, the twin daughters of Ghanaian emigrants living in Canada function as intercultural glitches; their treatment is symptomatic of their parents' cultural displacement. In an attempt to root themselves firmly within Canadian culture, the parents forego the ritualistic practices that the birth of twins would require among their family's community. When the girls start to exhibit challenging behaviour in a way that the author Edugyan clearly modelled on June and Jennifer

Gibbons (becoming increasingly insular, setting fire to neighbouring houses, etc.) the parents are nagged by the thought that this could have been avoided if they had followed the prescribed rituals. Like Jessamy in *The Icarus Girl*, the twins occupy a point of intersection between opposing epistemologies and have the dual function of binding the family to their cultural 'roots' while modelling emergent behaviour in the diaspora (the girls exhibit hyper-intelligence, the promise of which is neglected by the discriminatory treatment they receive from the community). This dual function of twinship is echoed in Buchi Emecheta's 1994 novel *Kehinde* which focuses on a first-generation Nigerian immigrant living in London who is haunted by the spirit of her twin who died during birth. At first, her twin seems to embody an uncanny 'return' of her Yoruba heritage that is being repressed by the requirement of assimilation to British life. When her husband forces her to have an abortion, the voice of her twin warns her: 'Our mother died having you. I too died so you could live. Are you now going to kill your child before he has a chance of life?'[80] The notion that she harbours within herself the spirit of Taiwo, her deceased twin, seems to render her incompatible with the requirements of an individualistic society. But the trajectory traced by the novel leads Kehinde towards a reconciliation, a middle path between the two systems. The achievement of this new diasporic existence is sealed through a union with her twin. '"Now we are one," the living Kehinde said to the spirit of her long dead Taiwo.'[81]

The use of twinship to forge practices of globalism at the intersections of colonial and digital network paradigms is clear in a documentary that explores concepts central to Afrofuturist aesthetics. *The Bight of the Twin* (2016), directed by Hazel Hill McCarthy III, follows musician and performance artist Genesis Breyer P-Orridge (best known for his work with the industrial music group Throbbing Gristle and psychedelic collective Psychic TV) as they mourn the death of their partner and fellow performer Lady Jaye. After they met in 1994, the pair embarked on what they described as a collective process of transformation. Inspired by the prominence of twins within the myth systems of vodou and santeria (of which Lady Jaye was a keen devotee) they began to manipulate their bodies through a series of surgical procedures with the aim of eventually resembling each other as closely as identical twins. The result of this process, which they described as 'pandrogyny', would be the production of a 'third entity' reassembled from their component parts. If twinship was the projected ideal of this process of mutual becoming, then the procedural model was provided by the cut-up techniques pioneered by William Burroughs and Brion Gysin: 'cutting up our bodies and our identities so that we become a third being'. Following Lady Jaye's death in 2007, Genesis became increasingly interested in the ritualist treatment of dead twins in West Africa. If Lady Jaye had become their twin through the process of pandrogyny, could he incorporate her spirit into his own after her death?

The film focuses on a journey undertaken by Genesis to Ouidah to Benin, a mythical heartland of vodou, which has 'the highest twin birth rate in the world' and where 'twins are venerated as Gods'. In a voiceover that echoes that of Martin Sheen in *Apocalypse Now*, Genesis's account of the trip evokes the trope of the colonial journey to the 'heart of darkness': 'We're going to go to Africa and find out where it [vodou and santeria] all began. The further back we can go the more we can find out what the connections are. Where do these come from – all these stories and beliefs?' Here, Genesis echoes the well-worn colonial discursive strategy of mapping the geographic otherness of Africa onto the temporal past. This use of a colonial motif is seemingly confirmed by the epigraph quotation from an 'Old Sea Shanty' that provides the inspiration for the film's title: 'Beware, beware, the bight of Benin for few come out though many go in.' However, the narrative of an encounter with otherness and the unexpected is undercut in the film by a counter-narrative that emphasizes repetition and the fact that, as Genesis explains in a phrase that itself is repeated so that it bookends the documentary, 'we found what we were expecting in Africa – a reinforcement – a vindication of something that we had built through our researches'. The journey was not a meeting with otherness, but a 'consecration of what we already expected. Or what we decided was going on in terms of consciousness and existence'. In other words, the twinship rituals in vodou are presented as just another commercial option in the global marketplace of spiritual remedies. Wandering around Ouidah with an ere ibeji statue slung around his neck, Genesis is presented as a tourist.

However, while *Bight of the Twin* seems to reinforce a colonial production of global space, its use of twins also complicates this interpretation. In a final voiceover speech, Genesis links vodou practices to both connectivity and mutation. 'Vodou is the earliest system trying to organise our thoughts along that possibility – that we are connected to everything'. Rather than a set of beliefs, vodou is presented as an information system in which the Orishas are 'midway creatures' that function as 'messages between the ultimate Gods and human beings'. The process of connection in vodou, Genesis emphasizes, is transformative. The encounter with the twin priestess, he explains, has engendered a 'mutation of his spiritual skeleton [that has] uncovered an inner structure [...] an indestructible hybrid'. In the last line of the film, following his speech about the bodily and spiritual mutation he underwent as a consequence of the project, he says: 'and this is when it swapped from being a documentary about vodou to becoming actually vodou'. This final statement enacts a rhetorical conflation between the audio-visual medium of the film and his own body that is entangled with that of his 'twin' Lady Jaye and the global vodou network. Here, twins are once again presented as a form of weird mediation that is used in the film to map a post-gender manipulation of the body onto a post-colonial diasporic global space that is encoded within vodou and santeria. Twinship, in other words, is employed to think transversally across and between these two widely diverging scales.

Bight of the Twin performs a temporality characteristic of what Eshun describes as the 'chronopolitical act' of Afrofuturism. Just as the trope of the twin is caught in a tension between the colonial uncanny and the transcorporeal weird, the film is a melancholic act of failed mourning (helping Genesis keep his dead loved one alive) and a proleptic act that transforms the viewer. On the one hand, twinship stands for the assertion of the past in the present through the state of melancholy. The ere ibeji doll strapped around the neck is a physical manifestation of Genesis's unwillingness to let go of his grief. But, on the other hand, twins are associated in the film with divination, the ability to read and therefore shape the future.[82] In this way, it echoes the way in which Les Twins performances act as points of intersection between the trauma of black historical experience and digital networks. Eshun dismisses the argument that Afrofuturism's focus on the future carries out a cultural betrayal by glossing over the traumas of the past. Rather, he argues that the work of Afrofuturists actually complements that of critics who have sought to 'assemble countermemories that contest the colonial archive' by 'situating the collective trauma of slavery as the founding moment of modernity'.[83] Afrofuturism 'extend[s] that tradition by reorienting the intercultural vectors of black Atlantic temporality towards the proleptic as much as the retrospective'.[84] The texts and performances explored in this chapter harness the looping temporality encoded within twinship to enact this double movement.

Notes

1   The genre of Afrofuturism has also framed the critical reception of Peskine's work. The 'Aljana Moon' portraits were exhibited as part of an Afrofuturism exhibition in Chicago's Museum of Contemporary Photography, while a short film that uses the

images as a starting point was screened in the 2018 Africa in Motion Film Festival as part of an event titled 'Decolonising the Speculative: Journeys to African Futures'.

2   Hillel Schwartz, *The Culture of the Copy: Striking Likenesses, Unseasonable Fascimiles* (New York: Zone Books, 1996), 39.
3   Robert Stam, 'From Hybridity to the Aesthetics of Garbage', *Social Identities* 3:2 (1997): 279.
4   Kodwo Eshun, 'Further Considerations on Afrofuturism', *The New Centennial Review* 3:2 (2003): 292.
5   Karen Dillon, *The Spectacle of Twins in American Literature and Popular Culture* (Jefferson, NC: McFarland & Company, Inc., 2018), 13.
6   Dillon, *The Spectacle of Twins*, 71.
7   Dillon, *The Spectacle of Twins*, 69.
8   Dillon, *The Spectacle of Twins*, 83.
9   Ian Baucom, *Specters of the Atlantic: Finance Capital, Slavery, and the Philosophy of History* (Durham, NC: Duke University Press, 2005), 18.
10 Elisha P. Renne and Misty L. Bastian, 'Reviewing Twinship in Africa', *Ethnology* 40:1 (2001): 8.
11 Misty L. Bastian, '"The Demon Superstition": Abominable Twins and Mission Culture in Onitsha History', *Ethnology* 40:1 (2001): 13.
12 Chinua Achebe, *Things Fall Apart* (London: Penguin, 2010 [1958]), 137–8.
13 Achebe, *Things Fall Apart*, 139 and 142–3.
14 Achebe, *Things Fall Apart*, 146.
15 Nick Briz, 'Thoughts on Glitch[Art]v2.0', http://nickbriz.com/thoughtsonglitchart/.
16 Briz, 'Thoughts on Glitch[Art]v2.0'.
17 J. Griffith Rollefson, 'The "Robot Voodoo Power" Thesis: Afrofuturism and Anti-Anti-Essentialism from Sun Rato Kool Keith', *Black Music Research Journal* 28:1 (2008): 85.
18 Helen Oyeyemi, *The Icarus Girl* (London: Bloomsbury, 2005), 210 and 256.
19 Babatunde Lawal, 'Sustaining Oneness in Their Twoness: Poetics of Twin Figures (Ère Ìbejì) among the Yoruba', in *Twins in African and Diaspora Cultures: Double Trouble, Twice Blessed*, ed. Philip M. Peek (Bloomington and Indianapolis: Indiana University Press, 2011), 85.
20 Helen Cousins, 'Helen Oyeyemi and the Yoruba Gothic: White is for Witching', *The Journal of Commonwealth Literature* 47:1 (2012): 47–58.
21 Homi Bhabha, *The Location of Culture* (London and New York: Routledge, 1994), 166.
22 Brenda Cooper, 'Diaspora, Gender and Identity: Twinning in Three Diasporic Novels', *English Academy Review* 25:1 (2008): 51.
23 Oyeyemi, *The Icarus Girl*, 210.
24 Oyeyemi, *The Icarus Girl*, 174.

25 The fact that Achebe's canonical novel is a key intertext is acknowledged directly when Jess and TillyTilly find a copy of the book, alongside Wole Soyinka's *A Dance of the Forests* (1963), in their grandfather's bookshelf.
26 Oyeyemi, *The Icarus Girl*, 66.
27 Diana Adesola Mafe, 'Ghostly Girls in the "Eerie Bush": Helen Oyeyemi's *The Icarus Girl* as Postcolonial Female Gothic Fiction', *Research in African Literatures* 43:3 (2012): 32.
28 Oyeyemi, *The Icarus Girl*, 235.
29 Oyeyemi, *The Icarus Girl*, 236.
30 Oyeyemi, *The Icarus Girl*, 239.
31 W.J.T. Mitchell, *Seeing through Race* (Cambridge, MA: Harvard University Press, 2012), xii and 5.
32 Mitchell adapts this definition from the work of Raymond Williams. See 'Addressing Media,' *MediaTropes eJournal* 1 (2008): 3.
33 Oyeyemi, *The Icarus Girl*, 86.
34 Dillon, *The Spectacle of Twins*, 85.
35 Dillon, *The Spectacle of Twins*, 86.
36 Marjorie Wallace, *The Silent Twins* (London: Vintage, 1996), 18 and 103.
37 Mitchell, *Seeing Through Race*, 40.
38 Tade Thompson, *The Rosewater Insurrection* (London: Orbit, 2019), 43.
39 Thompson, *The Rosewater Insurrection*, 44.
40 Thompson, *The Rosewater Insurrection*, 190.
41 Thompson, *The Rosewater Insurrection*, 191.
42 Thompson, *The Rosewater Insurrection*, 195.
43 Thompson, *The Rosewater Insurrection*, 159.
44 Thompson, *The Rosewater Insurrection*, 362.
45 Thompson, *The Rosewater Insurrection*, 362.
46 Thompson, *The Rosewater Insurrection*, 106.
47 Steven Shaviro, 'Review of *Rosewater* (2016) by Tade Thompson', *The Pinocchio Theory*, 16 January 2017, http://www.shaviro.com/Blog/?p=1422.
48 Zakiyyah Iman Jackson, *Becoming Human: Matter and Meaning in an Antiblack World* (New York: NYU Press, 2020), 1.
49 Thompson, *The Rosewater Insurrection*, 224.
50 Thompson, *The Rosewater Insurrection*, 73.
51 Jackson, *Becoming Human*, 2.
52 Dillon, *The Spectacle of Twins*, 12.
53 Joanne Martell, *Millie-Christine: Fearfully and Wonderfully Made* (Winston-Salem, NC: John F. Blair Publishing, 2000), 12.
54 Martell, *Millie-Christine*, 8.
55 Martell, *Millie-Christine*, 16.

56 Jackson, *Becoming Human*, 23.
57 Martell, *Millie-Christine*, 158.
58 Martell, *Millie-Christine*, 158.
59 Sianne Ngai, *Ugly Feelings* (Cambridge, MA and London: Harvard University Press, 2005), 7.
60 Birgit Abels, 'A Poetics of Dwelling with Music and Dance: Le hip hop as Homing Practice', *The World of Music* 8:1 (2019): 49–50.
61 Abels, 'A Poetics of Dwelling with Music and Dance', 51.
62 Abels, 'A Poetics of Dwelling with Music and Dance', 59.
63 Erin Manning, *Always More than One: Individuation's Dance* (Durham, NC: Duke University Press, 2013), 104.
64 W.E.B. DuBois, *The Souls of Black Folk* (Mineola, NY: Dover Publications, 1994), 3.
65 Elizabeth Alexander, '"Can you be Black and Look at This?" Reading the Rodney King Video(s)', *Public Culture* 7:1 (1994): 80.
66 6LACK, 'Never Know,' *Free 6lack*, LVRN and Interscope, 18 November 2016.
67 Fred Moten, *In the Break: The Aesthetics of the Black Radical Tradition* (Minneapolis and London: University of Minnesota Press, 2003), 5.
68 Moten, *In the Break*, 14.
69 Moten, *In the Break*, 18.
70 Moten, *In the Break*, 18.
71 Mark Butler, *Unlocking the Groove: Rhythm, Meter, and Musical Design in Electronic Dance Music* (Bloomington: Indiana University Press, 2006), 72.
72 Hiroki Azuma, *Otaku: Japan's Database Animals* (Minneapolis: University of Minnesota Press, 2009).
73 tobias c. van Veen, 'Vessels of Transfer: Allegories of Afrofuturism in Jeff Mills and Janelle Monáe', *Dancecult: Journal of Electronic Dance Music Culture* 5:2 (2013): 23–4.
74 van Veen, 'Vessels of Transfer', 24.
75 van Veen, 'Vessels of Transfer', 86.
76 Abels, 'A Poetics of Dwelling with Music and Dance', 49.
77 Reynaldo Anderson and Charles E. Jones, 'Introduction: The Rise of Astro-Blackness', in *Afrofuturism 2.0: The Rise of Astro-Blackness* (New York: Lexington Books, 2016), viii.
78 tobias c. van Veen and Reynaldo Anderson, 'Future Movements: Black Lives, Black Politics, Black Futures: An Introduction', *TOPIA: Canadian Journal of Cultural Studies* 39 (2018): 12.
79 Nicholas Royle, *The Uncanny* (Manchester: Manchester University Press, 2003). See also Roger Luckhurst, *The Invention of Telepathy 1870–1901* (Oxford: Oxford University Press, 2007), 148–80.
80 Buchi Emecheta, *Kehinde* (Oxford: Heinemann, 1994), 17.

81 Emecheta, *Kehinde*, 141.
82 Philip Peek describes various West African spiritual practices that associate twins with divination. 'Senufo diviners,' for instance, 'seek out twins in the other world with whom to communicate during their oracular sessions.' 'Introduction: Beginning to Rethink Twins,' in *Twins in African and Diaspora Cultures: Double Trouble, Twice Blessed*, ed. Philip M. Peek (Bloomington and Indianapolis: Indiana University Press, 2011), 11. Likewise, Ron Eglash traces a connection between the motif of doubling, twinning and divination. In a way that resonates with the works examined in this chapter he likens these divination systems to the recursive logic of cybernetics. *African Fractals: Modern Computing and Indigenous Design* (New Brunswick and London: Rutgers University Press, 1999), 89.
83 Eshun, 'Further Considerations on Afrofuturism', 288.
84 Eshun, 'Further Considerations on Afrofuturism', 289.

5

Twin faces as glitches in algorithmic image cultures

In November 2017, a number of identical twin social media stars with vast online followings posted videos on their YouTube accounts to mark the release of the iPhone X. There was one feature of the new range of phones that the twins were especially excited to test for their millions of followers: Face ID. The new biometric scanner built into the front camera system was being presented by Apple as the future of smartphone security, set to replace Touch ID on all subsequent models. The identification system's TrueDepth camera works by projecting 30,000 infrared dots onto the operator's face, which are used to trace its contours and produce a map. This map is then converted into a 2D image which becomes that iPhone's master key, without which it cannot be unlocked. This new security system seemed to be a gift to twin consumer tech reviewers determined to identify flaws in the interface: what better test for face recognition algorithms than identical twin faces? Most videos showed the twins were able to unlock each other's phones. As one of the Dolan Twins told their more than 9 million subscribers 'according to Apple, Ethan and I are the same person'.[1] The appeal of twin YouTube stars echoes what Lisa Zunshine calls 'cognitively enjoyable' exercises of the twin plots employed in early modern drama.[2] The mistaken identity trope produced through the twin characters in *The Comedy of Errors*, for instance, challenges the audience's essentialist biases by momentarily troubling the borders between individual identities and threatening to sever the bond, fundamental to capitalist modernity, between individual bodies and individual identities. In the Face ID videos, however, it is not just the cultural assumptions of audience members that are put to the test, but also those of the biotechnical systems of control that are becoming increasingly naturalized through their incorporation into everyday communication technologies.

Kelly Gates argues that the roots of our current faith in facial recognition technology, evidenced by the techno-utopian iPhone X marketing discourse, can be found in the uncertainty of the aftermath of 9/11. Following the attacks,

CCTV footage that emerged of two alleged hijackers passing through airport security prompted a claim that an automated rather than human-controlled face recognition system would have connected the images of the two men to their profiles in the CIA database and potentially prevented the catastrophe. The glitch in the system was not the technology but the human operator. This, Gates claims, is the origin story that has served to legitimize all subsequent facial recognition technology initiatives, instilling faith in greater automation and increasingly complex integration between bodies and information networks. As Lila Lee-Morrison puts it, biometric systems, 'as risk mitigation and security technologies, have specific desired outcomes, namely, to establish identity as clearly as possible in a geopolitical landscape of uncertainty and in the context of border control'.[3]

But Face ID is only the latest biometric technology promising to 'stabilize the messy ambiguity of identity, to automatically read a stable, individual identity off the body'.[4] Biometric systems emerged in parallel with twin studies methodologies during the second half of the nineteenth century, in response to the rapid growth of an urban population deemed a potential threat to security. Systems such as Alphonse Bertillon's anthropometrics, which combined photographic images with databases of information about body types to aid the identification of criminals, promised to render each individual of the growing urban mass identifiable and therefore the object of State control. The advantage of face recognition systems over other biometric technologies, such as fingerprint or iris recognition, is that they function at a distance and can therefore work without the consent of whoever is being scanned. Facial recognition technologies currently constitute the primary means of shoring up the connections between bodies and identities in a context in which the latter are increasingly produced by networked digital information.

Twin studies have always been closely entangled with biometric technologies, at the levels of both ideology and methodology. As William Viney points out, the twins used in Galton's experiments of the 1880s 'formed part of a wider perceptual apparatus; they gave a new way of seeing'.[5] Galton's interest in twins as evidence of the influence of inheritance on human nature is closely connected to experiments with photographic technology. Galton became interested in the potential of photography to visualize human types. For Galton, twin bodies and photographic technologies both held the promise of revealing hidden secrets about individual identity. However, the twin bond has also been used to question the effectiveness of these technologies and test their limits. A 2018 National Institute of Standards of Technology (NIST) report on the efficacy of commercial face recognition algorithms identifies twin faces as an enduring stumbling block. 'One component of the residual errors is that which arises from incorrect association of twins.'[6] Of the 127 algorithms tested, only one 'can correctly distinguish twins'.[7] However, as Kevin Bowyer and Patrick Flynn point out, in these tests the faces of monozygotic twins serve as adversarial images: training tools used to improve the capabilities of the technology. Furthermore, just as the 'exotic' appeal of twins was exploited as a marketing tool by the genetic studies experiments of the 1990s (such as the Minnesota Study of Twins Reared Apart), security software developers use twin tests as a publicity strategy.[8]

Despite the fact that simply by using these applications, the twin influencers are presenting them as desirable, by pointing to and embodying a failure in the ever more pervasive biometric systems, the twin Face ID videos open up a space for their critique. In his work on computer worms and viruses, Jussi Parikka argues that digital technologies that have been naturalized to the point of 'ontological invisibility' only reveal themselves in the event of breaking down.[9] Every media ecology, he argues, 'seems to have an accident of its own'.[10] Accidents 'reveal technology, and the power/knowledge relations that media are embedded in'.[11] Often described as genetic 'accidents', twins introduce mistakes or glitches into the dominant regimes of the power regulating digital cultures. Twin faces function as blockages in the increasingly smooth biometric interfaces that are binding humans with computational systems in more and more elaborate ways. In the process, despite their apparent frivolity, twin Face ID videos invite us to think the human-computer interface differently.

In this chapter, I examine how twins introduce blockages into biometric systems, in particular facial recognition technologies. I explore how these blockages constitute moments of uncertainty in the datafication of identity through a study of the performance of identical twinship as a form of strategic invisibility on social media in the context of biometric surveillance. The uncertainty of these events is echoed by other uses of the trope of twinship as a method for navigating algorithmic image cultures, including the location of 'twin strangers' on online image databases and the practice of the 'twin selfie'. These deployments of twins clash with the dominant social media logic of homophily which consolidates social divisions through a production of sameness and stasis. Unlike homophily, twinning here is characterized by a dynamic disequilibrium.

Misrecognitions of facial recognition technology

The videos in which twin influencers explore the limitations of commercial facial recognition technologies expose the contradictions that are characteristic of social media celebrity. They simultaneously present themselves as autonomous individuals, resourcefully carving out their own brand, and relays in information networks, mere conduits between commercial products and potential consumers. The hesitation between these two modes encodes wider tensions within digital image cultures, between a celebration and displacement of images and between conflicting regimes of power. As I argued in the Introduction, influencer twins such as the Dolan brothers and Brooklyn and Bailey present themselves as the human embodiment of technological connection. The electronic communications devices they use in their videos are presented as merely reinforcing the 'natural' connectedness that exists between them as twins. However, alongside this enactment of posthuman digital connectivity, in a video in which they test Apple's Face ID technology, the Dolan twins also perform their detachment from technology. Through their critique of Apple's new security system, they mark themselves as separate from and suspicious of the introduction of new systems of mediation. The premise and tone of the video parodies the claims to scientific objectivity made on behalf of the new Apple iPhone software. The twins carry out a series of experiments that test the limitations of the software in a way that echoes the studies surveyed by Bowyer and Flynn. Grayson and Ethan both in turn hold the phone at arm's length with a neutral facial expression, presenting themselves in false solemnity as species for the inspection of machine vision. A split screen technique which shows us the brothers gazing into their phones on one side of the screen and then the camera's perspective on their faces on the other side. [See Figure 5.1]

Figure 5.1 Grayson and Ethan Dolan trick Apple's Face ID security system in 'Twins Vs. iPhone X Face ID' by Dolan Twins, YouTube, 4 November 2017. Available online: https://www.youtube.com/watch?v=GFtOaupYxq4 (accessed 7 July 2021).

These tensions are reproduced in the physical act of interfacing or attempting to interface both with their electronic devices and with each other. While the mock objectivity of the influencers performs the scientific detachment of the mug shot, by operating the cameras themselves, the Dolans, like all Face ID users, are both subject and object of the disciplinary gaze, a coincidence of roles that blurs the 'see/being seen dyad' that Foucault places at the heart of the disciplinary apparatus.[12] Furthermore, the act of extending the arm to take a photo of yourself with a smartphone has become a visual shorthand for the extent to which bodies adapt to the affordances of our technological devices. Unlike the arrangement used in mug shots, the camera is incorporated into the body, a part of it rather than an alien entity. The Dolans alternate between emphasizing the distinctness of their individual faces ('If you look really closely, they're a little different') and playing up to their similarities (they frame them in the same way using identical hats). On the one hand, part of the star appeal of the Dolans is that only their followers, those who really know them, can tell them apart with ease. This dynamic reinforces the myths of individuality and is part of the process of 'self-branding' that is central to the digital labour of influencers.[13] On the other hand, they triumph in being mistaken for 'biometric doubles'. The apparent repetition between Ethan and Grayson's faces pre-empts the repetition produced in the act of photographic capture by the Face ID system, a doubling that is emphasized by the use of split screen technique.

By foregrounding this hesitation, the twin Face ID videos evidence a tension at the heart of algorithmic image cultures more generally. They perform the paradoxes engendered by what Daniela Agostinho describes as the 'datafication of vision'.[14] Since 'machine vision occurs through data, not optical means [...] vision becomes essentially post-optical'.[15] And yet images and metaphors that evoke vision are everywhere. As vision becomes datafied, 'optics is both displaced and reinstated'.[16] This shift in the ontology of the photograph is reflected by its role in image-sharing social media platforms. In many ways, social media has been a vehicle for the growing influence of images in social life so that, as Tama Leaver et al. put it, the 'material world has sought to become "Insta-worthy" in redesigning practices, cultural institutions and material spaces'.[17] However, in social media photography, the image has become computational, marking a shift away from the importance of the visual. The computational nature of the image is foregrounded with particular clarity by the photo-sharing app Instagram, which positions the metadata produced in the acts of photographic capture and distribution – from tagging to recording the time and place of upload – as being equally if not more important than the content of the images themselves. Furthermore, in the transition from taking a photograph and uploading it to your profile, the manipulation of the image is a necessary stage that cannot be bypassed. In Instagram, as in computational culture more generally, there is no opting out of either 'filtering' or 'editing' the image. Alexis Madrigal points out how, since the beginning of the smartphone era, apps have offered to 'upgrade your face' by using algorithmic systems to produce an image of the face that statistically the user is likely to want to see.[18] However, unlike previous models that had a 'flaw-eliminating beauty mode' that you could turn on or off, the new iPhone XS range makes this 'skin-smoothing' feature a default. Since the photographic demands of selfie culture are a major driving force behind the technological innovations of phone manufacturers and app developers, the 'global economy is wire up to your face'.[19]

The result of this paradox, according to Daniel Rubinstein and Katrina Sluis, is that the networked image is characterized by the co-existence of two seemingly incompatible visual logics: on the one hand, a 'rational, visual representational logic according to which the image on our screen refers to a cat somewhere in the universe' and, on the other, a 'recursive, viral logic of intensity, multiplicity and incompleteness in which the image refers only to itself'.[20] While photography retains an aura of representationalism and is made to speak the language of identity, as digitization 'breaks the chain of signifiers' it also creates 'an excess' or 'supplement' that is 'not representational but sensual and affective'.[21] This double logic within algorithmic image cultures serves to reinforce historical identity categories while simultaneously binding us to algorithmic informational systems at a pre-individual level. Through their exploration of the iPhone X's face recognition system, the social media twins draw attention to the paradox of algorithmic image cultures. The seemingly double image of the Dolan Twins' faces seems to pre-empt the proliferating, decentred logic of the image in networked cultures, the simultaneous appearance of the same image across multiple platforms. The appearance of Ethan and Grayson's faces next to a split screen containing one sibling's face within the Face ID interface presents their twinship as an embodiment of this proliferating logic. Yet, on the other hand, the twins present their faces as an obstacle to machine vision, a glitch or stumbling block in the integration of human life and algorithmic identification processes, indexing a decoupling of the image from its referent. If one sibling's face can access either of the twins' digital identities, then it ceases to be representative.

The tension between two visual logics within the Dolans' twin Face ID video foregrounds how facial recognition systems mediate between distinct systems of power. In many ways, face recognition systems are exemplary of the emerging systems of power in digital cultures that function through a logic of market-driven modulation or control. Biometric technologies are the most efficient interfaces with digital networks and serve to embed the logic of information capitalism within the body itself. Combined with automated facial expression analysis, face recognition technology generates the type of information necessary to the efficient functioning of a control society: information about consumer habits, patterns of spending and debt, geographical movement and the consequent affective modulations of the body. Markets have the capacity to identify patterns in this data and cater to events, crises or desires (whether individual or transindividual) before they arise. Faces in biometric systems are reduced to patterns of digitized information (produced by the 'mapping' with infrared dots) and function not as windows onto the essential identity of the individual operator but as a point of access to information networks. As James Ash et al. point out, the 'smoothness' of the interface between individuals and computer systems enabled by facial recognition technologies would enable the 'open circulations' that, according to Deleuze in his influential 'Postscript on the Societies of Control', replace the closed milieus of disciplinary power and society.[22]

However, central to the success of face recognition systems is their simultaneous appeal to both disciplinary and modulatory regimes of power and the visual logics that go with them. They seem at once extensions of criminal identification strategies rooted in the nineteenth century and emblematic of the emerging regime of modulation. As Sarah Kember puts it, facial recognition technologies 're-produce the norms of nineteenth-century disciplinary photography even as photography becomes allied to the security-based biopolitics of computational vision and smart algorithmic sorting'.[23] Recent media studies research has sought to understand why network culture, rather than ushering in a postracial or postidentitarian era, has fostered a rise in identity politics. Wendy Chun uses the term 'homophily' – which is Greek for love as love of the same – to describe a logic that fuels and justifies online discrimination that reproduces, entrenches and legitimizes existing forms of social segregation. This logic functions not just at the level of the echo chamber – the idea that you find what you are looking for on the internet and have your previously held opinions or prejudices confirmed by the opinions of others – but also at the level of what Chun calls 'pattern discrimination', the shoehorning of data produced by online analytics software into simplistic identity categories that reproduce 'older' racial and class categories.[24] These algorithms, according to Chun, do not only capture the social identities that they find online but rather they actively produce these identities;

they are not only 'descriptive but also prescriptive and performative'.[25] In other words, the algorithmic logics that subtend social media, and which are facilitated by biometric interfaces, reproduce historical identity categories that legitimize social inequalities through a focus on individual habits and preferences as these are expressed through online activity.[26] Chun's concept of 'homophily' is echoed by the work of a number of researchers who have explored the ways in which algorithmic systems, including biometric technologies, reinforce historical racism. Simone Browne, for example, has used the term 'digital epidermalization' to describe the role of biometric technologies in the reaffirmation of the 'visual economy'[27] of race at a cultural moment when the focus on the body at the molecular level might provide the conditions for what Paul Gilroy (2000) termed a 'postracial humanism'.[28] The paradox identified by Chun is present at the level of the interface through the way that Face ID (and similar systems) converts the human face to networkable digital information while reproducing the visual conventions of nineteenth-century ethnographic photography.

By performing a moment of failure, the twin Face ID videos have an unintended critical effect comparable to that of a number of artistic works that use facial recognition technologies. In her study of interventions by Thomas Ruff, Trevor Paglen and others, Lee-Morrison argues that art that engages with biometric systems does two things. On the one hand, it can 'provide a cultural translation of the technology' by placing it within the historical and political contexts that are occluding by tech marketing language. On the other hand, by 'decontextualising its processes' these artworks 'allow for an engagement with this technology that not only problematizes its use but also imagines alternative outcomes of the technology and its processes'.[29] The videos by twin influencers discussed here carry out both of these gestures: they place facial recognition technologies within a fragile point of intersection between contradictory systems of representation and power while pointing to emerging positionalities at the borders of the visible.

Loss of face

The paradoxes of the algorithmic image (its proliferation belying a shift away from the visual; its simultaneous affirmation and unseating of the logic of representation) are mirrored by the role of the face in digital culture. The omnipresence of faces in networked culture coincides with a collective loss of faith in the visage as a seat of subjectivity. Claire Colebook argues that the technology-driven emphasis on instantaneity, disconnected emotions and flashing screens is rapidly producing what she describes as a 'total loss of face' in contemporary culture.[30] The 'face' that is being lost, according to Colebrook, is that which was described by Emmanuel Levinas in his ethical philosophy. Faces, according to Levinas, constitute the source and provide the motivation for ethical behaviour since they provoke an 'awakening to the precariousness of the other', which serves as a reminder of our own state of vulnerability in the world and hence our responsibility towards one another.[31] As Colebrook explains, Levinas's concept of the face 'relies on a singularity that would be liberated from all generality, that would not be a specification of this or that universal type'.[32] In a social world of over-exposure to photographic depictions of others, in which we constantly swipe away faces on social media and dating apps, visages have lost their specificity and, along with it, their supposed connection to an interior reservoir of humanity. Ironically, Facebook heralded the demise of the face at the very moment of its seeming triumph. For Colebrook, one of the most visible signs of this cultural loss is the rise of the smiley face. 'So lacking in distinction that it has neither race, nor humanity, nor artfulness, the smiley face signals loss of life'.[33] No matter how many iterations and variations are produced by enterprising software companies, the endlessly serialized smiley face marks a 'retreat from specification and the removal of any definitive body – anything that would allow for engaged sympathy'.[34]

Twin faces proliferate within this seemingly paradoxical coincidence between the omnipresence of faces and a cultural 'loss of face'. A growing number of apps and websites are offering to find users' 'twins' by scouring the internet for similar faces. The website twinstrangers.net, which, at the time of writing, has more than 8.5 million paying subscribers, claims to be able to 'find your lookalike from anywhere in the world'. A video uploaded to YouTube explains that 'when you register, our AI face recognition instantly compares your photo against millions in our database finding your closest matches'. If two subscribers are matched they have the option of contacting each other. In one of the first promotional videos for the site, viewed more than 9 million times on YouTube, Niamh tracks down her 'doppelgänger' Karen using the Twin Strangers software. Two follow-up videos show Niamh meeting her second and third twin strangers. During her second trip to Italy, Niamh is told that she not only resembles her lookalike Luisa but shares a very similar 'aura'. Hearing this news, Niamh breaks down in tears. The discourse surrounding Twin Strangers echoes the ambiguity of the twin social media stars. On the one hand, it seems to celebrate the cultural loss of face driven by networked image databases. In this respect, it echoes the appeal to the uncanniness of social media of the documentary *Twinsters*, discussed in the Introduction. The profiles we construct for ourselves online are at once intimately connected to our everyday experience and detached, like doppelgängers that might take on agency of their own. While social media users employ the same language, visual aesthetics and narrative techniques, these profiles become increasingly interchangeable. But on the other hand, it constructs the face as a site of potential ethical communication. The site promises not only to match lookalike users but also to facilitate a deep interpersonal connection.

Twin social media stars like the Dolan Twins and Brooklyn and Bailey mobilize their twinship in the name of a similar ambiguity. They revel in merging into oneness – they accentuate their similar appearance by speaking at the same time and wearing the same clothes – while encouraging their real followers to identify the differences between the siblings and know the 'real them'. They present themselves as literally embodying the tensions at the heart of what technology writer Jay Owens describes as the 'age of post-authenticity': 'a strange, surface-centric moment in popular, digital culture right now – where the original "essence of things" has indeed become somewhat unfashionable (or just less entertaining)'.[35] Just as successful memes produce 'unexpected, witty truths' out of 'the most inauthentic, borrowed, or stolen stock content possible', the aesthetic of YouTube stars is one of authentic inauthenticity. They are inauthentic in a way their followers can identify with. Another closely related aesthetic of social media is that embodied by the *Kinfolk* brand, which has diffused into general hipster culture. If teen YouTube stars are authentically inauthentic, then *Kinfolk* is inauthentically authentic. The paradox of the *Kinfolk* aesthetic is exposed most effectively by the Instagram parody account Sociality Barbie. With a tagline that reads 'Adventurer/Coffee Drinker/Authentic Living', the posts play on the tensions inherent to mass-produced and mass-consumed authenticity. Images of barbie dolls posing in front of wild landscapes alternate with close-ups of coffee mugs and *Kinfolk* magazines spread out on zinc tabletops. A video produced by *Kinfolk* to accompany its seventeenth issue on the theme of 'family' constructs a portrait of 'inseparable' Parisian twin sisters Monette and Mady who 'live in the same apartment, dress alike, dance together and share every meal'. In the clip, the sisters are made to embody the contradictions of the *Kinfolk* aesthetic. Dressed in identical red linen dresses and alone in a cavernous warehouse loft – that emblematic post-industrial space of hipster culture – the siblings dance synchronously against the sound of them musing about what it means to be a twin. 'To be a twin is provocative – it asks questions', one of them says, before the other completes the thought: 'Yes, it asks questions about singularity.' Through

Monette and Mady's performance of their entwined identity, twinship embodies the precarious status of individuality within 'inauthentically authentic' hipster culture. By embodying the contradictions within social media image cultures, twinship tests its limits, pushes it to breaking point.[36]

One of the effects of looking at photographic images of twins is that the faces staring back at you take on a mask-like quality. Rather than just the result of monozygosity, the impression of identicalness in many images of twins is the product of a performance. In photographic stagings of identical twin identities, the sitters employ a form of mask – the depths of emotional and biological differences are occluded by a surface sameness that is presented to the camera. In Peter Zelewski's 2018 photobook *Twins* the photographer and the photographic subjects collude in their performance of identicalness.[37] The supposed objective neutrality of the camera is pre-empted by the deadpan expressions of the twins themselves as they present themselves as specimens that are representative of the human type announced by the book's title. One of the reasons for this mask effect may be the fact that, with many monozygotic twins, facial expression and movement – the play of emotion through their features – often breaks the spell of identicalness. My daughters, for instance, have always looked most alike when their features are at rest. It is mainly that which animates the face – the choreography of the features – that produces the differences. In Audrey Niffenegger's body-switching twin novel, *Her Fearful Symmetry* (2009), when a ghost inhabits the body of one of the identical twin protagonists, her own sibling does not recognize her.[38] This seemingly farfetched plot contrivance is probably most believable to somebody who is close to twins themselves and knows the transformative effect of different facial animation. It is for this reason that twins posing as twins usually assume a deadpan expression, an attitude that evokes both the mug shot and face recognition scanning portraits.

Just as the machine vision of the Face ID system conflated Ethan and Grayson's faces, when the human eye confronts a photographic portrait of identical twins for the first time, it tends to move back and forth between the two faces in search of differences on which to anchor itself. The gaze skates across the surfaces of the faces presented to Zelewski's camera just as the eye slides off the surface of a mask during a Noh performance. It is significant that the Twin Strangers videos all end with Niamh and her lookalike putting on make-up in an attempt to accentuate their similarities. It is only when they are made-up in the same way and assume the same affectless expressions that the uncanny effect is realized. Twinship here is literally a mask. The mask-like effect of unknown twin faces evokes the use of actual masks by North American photographer Ralph Eugene

Meatyard in his project titled 'The Family Album of Lucybelle Crater'. The sixty-four-image series, which Meatyard completed in the two years before his death in 1972, is a sinister parody of a family album in which all the figures pictured, which include the photographer himself alongside friends and real family members, are wearing masks. Twinness is suggested both by the fact that only two masks are used throughout the series and by the repetition of a single name in the captions scrawled by hand beneath the images. (A typical caption reads, 'Lucybelle Crater and her 46-year-old husband Lucybelle Crater'.)[39]

In her interpretation of the series, Marianne Hirsch argues that Meatyard's use of masks is a grotesque, exaggerated performance of a quality common to all family photographs. In family photo albums, identity is presented not as something that is individual, but rather a quality that is 'defined by the mask of familial relation and of photographic convention'.[40] Meatyard's masks are a metaphor for the way we project particular ideological 'frames' or fantasies onto photographs. In the case of family albums, the fantasy is that of individual identity being determined and conditioned by belonging to a family. In contemporary digital cultures, the narrative function of the family photograph album has partially been replaced by family WhatsApp groups. WhatsApp groups have become the main platform through which family relations are performed, reasserted and naturalized through the sharing of photographic images. In many ways, these digital fora echo the main qualities of the analogue photograph album. The fact that they are private (only shared with family and friends) and separate from the general commercial circulation of images reaffirms an imagined separateness of the family unit. Family WhatsApp groups are often dominated by a desire for an equal distribution of images. Efforts will be made to avoid a single grandchild dominating the thread, for instance. The mask of family belonging in this case is imposed by the medium of sharing itself. The context of the family WhatsApp group means that the images that are shared within it are interpreted as images of familial belonging.

In some ways, the performance of twinship that you come across in many images of twins is merely an augmented echo of the performed nature of all family photography. However, in other ways the mask-like quality of twin faces disrupts the conventions of family photography. The ideological function of the family photo album depends upon the easy recognition of the separate identities of the various family members as well as their distinct individual placement within the family structure. Much of the pleasure of looking at old family photographs is derived from the satisfaction of identifying who is who and discerning their place within the wider, largely imagined picture of blood relations. This satisfaction ultimately serves the reaffirmation not only of the familial structure but also the more general patriarchal hierarchies that family both reproduces and naturalizes. Photographs of identical twins disrupt the easy recognition and allocation of the individual's place within the wider family hierarchy. When my daughters were babies, grandparents, uncles and aunts often had difficulties identifying who was who in WhatsApp photos. It was only after zooming in to an identifying feature that distinct identities could be reaffirmed. Even for parents, old photographs of twins as babies can often have a troubling dislocating effect as we struggle to tell who is who.

The way that identical twins can jar the dynamics of family photography is played out in a particularly sinister fashion in the television series *Big Little Lies* (2017). The HBO miniseries is set among a community of anxious and high-achieving young parents in Monterey, California. The ever-present tensions within the community start to intensify when a six-year-old girl in the local public school accuses a boy who is new to town of physically abusing her. Through seven episodes the intertwined storylines expose how the children become proxies for conflicts and desires among their parents. One set of parents, for whom conflict and desire are particularly intertwined are Celeste and Perry White, played by Nicole Kidman and Alexander Skarsgård. Rich, powerful and still seemingly passionately in lust despite having two young children, the Whites give every outward appearance of marital happiness. But beneath the photogenic exterior there is a violent reality of ongoing abuse. Celeste, struggling to come to terms with her situation, keeps the signs of abuse – bruises and black eyes – rigorously out of view with long-sleeved, high-necked tops and dark glasses. Through the Whites, *Big Little Lies* explores the ways that the myth of the family unit can act as a mask for the darkest of secrets.

Celeste and Perry's children are identical twin boys Max and Josh – played by Nicholas and Cameron Crovetti – and their twinship is used to emphasize the disparity between surface illusion and hidden realities. One of the first times we encounter the twins is when Celeste is taking photographs of them on the balcony of her minimalist and coldly beautiful house against the spectacular backdrop of the Pacific Ocean. When Perry returns from work, they both swipe through the photographs on her iPhone, deleting the pictures in which one of the boys has his eyes closed. The viewers are shown the editing process over her shoulder as Celeste digitally erases the blemishes from their collective appearance in an attempt to uphold the high aesthetic standards she projects into her family unit. In the photographs we struggle to tell the boys apart, just as they are treated as a single entity in the series itself right up until the penultimate episode when we learn that it was not the newcomer but one of the White boys, Max, who choked the girl in school. The plot twist is a key moment in the series since it reveals that much that had been thought of as hidden from view had been on display all along. Despite Perry's desperate attempts to keep his violence under wraps, his sons knew about it all along and have begun to copy their father's behaviour. But the revelation also plays on the very specific relationship between twins and the borders of visibility in an age of digital images. Because of their likeness and despite their constant exposure to Celeste's camera, the boys remain curiously obscured from view. The way they blend into one another in the viewer's mind shields them from being objects of individual scrutiny just as the faces of the Dolan Twins and Brooklyn ad Bailey evade Apple's biometric security systems.

The mask-like quality of twin faces in visual mass culture underscores the disruptive potential of the Face ID videos. Discussing state use of biometric systems during the Occupy movement, Zach Blas argues that the use of masks by protestors (including the iconic Guy Fawkes mask borrowed from David Lloyd and Alan Moore's graphic novel *V for Vendetta*) were 'forms of queer illegibility'.[41] These 'aesthetic and political practices of anti-normativity and anti-standardization' undermine neoliberal surveillance systems and create 'amorphous, encrypted, incalculable, excessive and weird collective stylings of bodies and environments'.[42] These 'queer biometric failures', Blas concludes, 'are utopian oppositions that do not cohere to state visualizations or representations; they evoke facelessness and defacement [...] and make the face a nexus of refusal, collectivization, and potentiality'.[43] In their Face ID videos, the twin social media stars enact moments of 'biometric failure'. By performing identicalness for the camera, they construct their faces as 'natural' masks that jar with State visualizations. In the process, and despite the apparent frivolity of the videos, they point to the critical potential of twinship in defamiliarizing the emerging identification systems of the digital age. Jenny Edkins identifies a contradiction at the heart of contemporary 'face politics'. 'With a shift from the modern episteme to a world of digital images' we might expect the face to be disappearing.[44] However, in social media cultures 'the face endures as an emblem of political personhood'.[45] In the face of this contradiction, Edkins outlines a political strategy of 'tarrying with the precarious existence of the face: the face neither mantled nor dismantled, neither sustained nor destroyed, but rather revealed as both there and not-there, here and not-here at the same time'.[46] The mobilization of twin faces as masks enacts a face that flickers between visibility and invisibility. In the process, the form of transfaciality performed by twin faces points to alternative configurations between faces and personhood.

Twin selfies and human-machine assemblages

Like most twin influencers, one of the stock visual genres employed by the Clermont twins, Shannon and Shannade, is the joint portrait using still photographic images. That many of these are self-portraits is made evident by the fact that the siblings are standing, smartphone in hand, in front of a mirror, or from what has become the defining characteristic of the selfie: the presence of an extended arm visible at the bottom of the frame. The #twinselfie both deploys and undermines the conventions of what is the most common vernacular photographic practice of the digital age, a convention that, according to Madrigal, is driving the increasingly complex use of algorithmic systems in smartphone cameras and image apps. On the one hand, the twins exploit the cultural associations between the genre of the selfie and authenticity. In contrast to some of the more staged and aesthetic photographs posted onto their profiles, whether they be model shots promoting their merchandise or the ubiquitous staged lifestyle images, the selfies are presented as unplanned and therefore windows onto the realities of their everyday lives. However, the selfies are exemplary of the paradoxically highly mediated immediacy of a mode of stardom that relies on an aesthetic of authentic inauthenticity. On the other hand, as evidenced by the oxymoronic nature of the hashtag #twinselfie, the genre undermines the other key cornerstone of the selfie: the focus on the individual self. The self that the twins present in their twin selfies is a mutually entwined self, a photographic performance of the dynamic of entwinement that is central to their brand.

The selfie has proved to be a particularly fruitful focus for the analysis of the socio-technical assemblages enabled by algorithmic photography. The photographic practice is often dismissively described in the popular press as being symptomatic of a narcissistic culture. It is easy to see how the boom of the selfie could be marshalled as evidence of a collective cultural self-obsession. The prevalence of this mode of expression seems to indicate the reduction of public discourse to what Zygmunt Bauman terms 'life politics' in which the subject, stripped of any real political agency and unable to connect his or her experience to that of the wider social field, is reduced to maintaining an illusion of agency through the exercise of market-driven lifestyle choices.[47] The association between twinship and incest in contemporary mass popular culture has also been described as indicative of cultural narcissism. Stephen Marche points out that, far from being taboo, incest has become a clichéd plot development in popular culture, a trope that is often played out through narratives of twins, whether they be Luke Skywalker and Princess Leia in the *Star Wars* franchise or Jaime and Cersei Lannister in *Game of Thrones*.[48] Marche argues that twin incest narratives reveal that incest is really an extension of narcissism. When Siegmund and Sieglinde fall in love in *Die Walkure*, they are really falling in love with their own reflected images. Like the boom of the selfie for many of its critics, for Marche the naturalization of the twin incest trope is indicative of societal atomization.

But rather than reaffirm the narcissistic qualities of the selfie, its appropriation by social media twins emphasizes the constitutive connection between self and information networks in digital cultures. Selfies have been attracting increasing critical attention for the light they shed on the complex forms of intersubjectivity of the digital age. Rather than reflective of the technology-driven atomization of communities, they are more often viewed by digital media scholars as socio-technological assemblages. Aaron Hess, for example, focuses on how selfies illuminate emerging and evolving 'relationships between technology, the self, materiality, and networks' of the digital age.[49] The 'selfie assemblage', as he terms it, gives expression to 'the affective tensions of networked identity: the longing for authenticity through digitality, the conflicted need for fleeting connection with others, the compulsion to document ourselves in spaces and places, and the relational intimacy found with our devices'.[50] Selfies both 'announce' these tensions and act as ways of 'coping' with or reconciling them. The connection between self and networks in the twin Face ID videos evidences these tensions between desire for authenticity and meaningful interpersonal connection and an ecstatic embrace of the reduction of life to information and the production of connective metadata.

Like Hess, Paul Frosh argues that selfies should not be understood purely in visual terms, but rather for the ways in which they integrate photographic images into 'a technocultural circuit of corporeal social energy' that he terms 'kinesthetic sociability'.[51] A key technological innovation that has enabled the selfie boom is the design of the smartphone. The fact that they can be easily held and operated with the same hand, that they display an image of the 'pre-photographic scene' large enough to be viewed at arm's length, and that they have lenses on both the front and back mean that, unlike with traditional camera design, smartphones no longer function as a 'barrier between visible photographed spaces and undepicted locations of photographing and viewing'.[52] The two main consequences of this are that 'the space of photographic production or enunciation is effortlessly unified with the space of the picture itself' and that 'the unified space of production and depiction becomes a field of embodied inhabitation' since the camera is 'literally incorporated, part of a hand-camera assemblage'.[53] Selfies mobilize the indexical aura surrounding photography as part of a 'connective performance' between gestural images and their habituated embodied responses (whether tapping the screen in approval or swiping in dismissal) rather than to authenticate 'semantic reference'.[54] Twin selfie photographs reproduce in microcosm the connections between bodies and technologies that constitute the 'kinesthetic sociability' of selfie assemblages. The genetic connections between each other's bodies, underscored by their performed twinship, pre-empts the techno-corporeal circuits into which they are woven by their embedded metadata.

Rather than a reaffirmation, the twin selfie is a challenge to the narcissism of the viewer. In this respect, the twin selfie exposes the fact that the fascination with twin YouTube stars bears an uncomfortable resemblance to the form of spectatorship encouraged by the staging of conjoined twins in the 'freak shows' of the nineteenth century and first half of the twentieth. One of the most memorable insights into these modes of display is provided by Tod Browning's 1932 film *Freaks*. Since the majority of the actors were real circus performers (including the torso-less Johnny Eck and the microcephalic Schlitzie) the movie constitutes a unique meeting point between early 1930s Hollywood and the North American sideshow tradition. Two of biggest attractions in the film were Violet and Daisy Hilton – the 'Hilton Sisters' – conjoined twins who by the time *Freaks* was filmed were household names in the United States. The drama and tension in the film turns on who holds the power to look at whom. The 'performers' are habituated to their usual status as the object of other people's gazes and, as such, they are accustomed to being subject to other people's power. The ending of the film stages an uprising in which the performers – who joyfully inhabit the label of 'freaks' – seize power by confronting a pair of normatively embodied performers who are trying to exploit them. The sequence of images that narrate this rebellion emphasizes the act of defiant looking back as the freaks direct their collective gaze at those deemed normal. Most of the shots in the latter stages of the film are of performers simply looking. The film narrativizes what Elizabeth Grosz describes as the 'intolerable ambiguity' of the spectacle of conjoined twinship. The normative function of contemplating conjoined bodies in a freakshow 'confirms the viewer as bounded, belonging to a "proper" social category'.[55] The viewer's horror, on the other hand, derives from 'recognition that this monstrous being is at the heart of his or her own identity, for it is all that must be ejected or abjected from self-image to make the bounded, category-obeying self possible'.[56] In a way that echoes the 'cognitively enjoyable games' described by Zunshine, the dual reaction to conjoinment described by Grosz sends a tremor through the spectator's narcissistic self-image of integrity. It is this tremor effect that the spectacle of social media twinship shares with the freakshow.

While the twin selfie confronts its viewer with the socio-technological assemblages that are constitutive of identity, it also foregrounds the face itself as an assemblage. While faces are actually assemblages of different elements (skin, muscle, cartilage, bone, affect, cognition, etc.), cultural practices such as portraiture have created the face as a single, unified object, detachable from its bearer and expressive of his or her essential identity. In an influential chapter of *A Thousand Plateaus* titled 'Year Zero: Faciality', Gilles Deleuze and Félix Guattari describe the often violent cultural imposition of unity on the facial assemblage as the 'abstract machine of faciality', which renders the face 'inhuman' by turning it into a 'phantom'.[57] Gates argues that face recognition systems are part of what Deleuze and Guattari describe as the 'technological trajectory' of the abstract machine of faciality. Despite the fact that, at the point of interface, systems such as Face ID reduce the face to an assemblage of contours, the social and political uses of the information produced by the interface reaffirm the process of facialization: the attachment of the image of a face to a stable individual identity. Twin faces constitute a challenge to this process. By setting the spectator's eye in motion – whether it be a human eye or that of biometric machine vision – they deterritorialize the face. The way the eye moves back and forth across twin faces frees their features from the false unity of facialization in a way that echoes Deleuze and Guattari's description of the processes of 'dismantling' that make 'faciality traits themselves finally elude the organization of the face – freckles dashing toward the horizon, hair carried off by the wind, eyes you traverse instead of seeing yourself in or gazing into those glum face-to-face encounters between signifying subjectivities'.[58] In this, they are anticipating an effect that Deleuze associates with the close-up in his later books on cinematic aesthetics. In a way that echoes this process of 'dismantling', cinematic close-ups of faces have the power to 'tear the image away from spatio-temporal co-ordinates in order to call forth the pure affect as the expressed'.[59]

Mark Hansen's concept of the Digital Facial Image is a useful point of reference for the role of twin faces in relation to the 'faciality' of biometric systems. Hansen uses the term to describe aesthetic experimentations with human-computer interfaces carried out by new media artists that focus on the human face. Kirsten Geisler's installation piece *Dream of Beauty 2.0* (1997), in which the user is invited to interact with the limited emotional and verbal range of a disembodied computer-generated face, is taken as exemplary of the way Digital Facial Images 'draw attention to the non-seamlessness of the interface between embodied human beings and the computer'.[60] Like Deleuze's close-up, experiments with the DFI 'strike against late capitalist semiotic mechanisms that function specifically by reducing embodied singularity to facialized generality'.[61] Unlike the close-up, in which the face functions as a vector of deterritorialization – 'a liberation of affect from the body' – the DFI acts as a 'catalyst for a dynamic re-embodiment of the interface'.[62] For Hansen, these experiments with what he calls the DFI hold a similar utopian potential to the mask in Blas's account. The 'interactive coupling' between humans and computer systems carried out by these aesthetic experimentations, 'catalyse the production' of 'new affective relations' that are emerging at the intersection between the 'expanded virtual domains' of digital technologies and the virtuality of the body. Writing in the early 2000s, Hansen could not anticipate the expansion of the range of 'interactive coupling' between humans and computers that would be naturalized by social media. The sheer banalization of these human-computer interfaces through their integration into everyday life inevitably detracts from the critical potential that Hansen associated with the Digital Facial Image. In many ways, social media has served to harness the potential of what Frosh called the 'kinesthetic sociability' of selfie networks to the development of 'brand assemblages' (a term used by Celia Lury).[63] The jarring genre of the twin selfie, I argue, restores some of this critical potential. The 'hesitation' that twin selfies introduce into selfie culture draws attention to the socio-technological assemblages that are constitutive of identity and irreducible to the imposition of identity categories carried out by algorithmic systems and their operative logic of 'pattern discrimination'.

The Clermont twins' articulation of the twin selfie with Afrofuturist aesthetics is particularly provocative. Since coming to fame through the 2015 reality TV show Bad Girls Club, Shannon and Shannade Clermont have developed a huge social media following and modelling career. Their aesthetic is dominated by two intertwining tropes: the symmetry of 'identical' twinship and a punk futurism expressed in silver lame clothing, robotic affectless facial expressions and conspicuous body modifications. In the video for Nicki Minaj's 2018 hit 'Good Form', the sisters featured as a pair of symmetrical android servants, dancing in the background and performing as human stools. A feature in *i-D* magazine from the same year was illustrated with photographs of the twins in long silver hair and outsized mirror shades striding across a futuristic desert landscape. The Afrofuturist aesthetic frames the way in which they navigate social media, underscoring the fact that their bodies are reconfigured by the affordances of network technologies.

In Chapter 4, I examined the ways in which trope of Afrofuturist twinning has been used to speculate about emerging transcorporeal configurations of information and media by artists and authors of speculative fiction and black futurism. Agostinho has identified Afrofuturism as a tool with which artists 'compel us to rethink and possibly expand the notion of visibility' in the context of datafication in which 'visibility as a regime of algorithmic structuring coexists with the notion of visibility as a political site for subject formation, justice claims and resistance'.[64] The Clermont sisters' Afrofuturist twin selfies are strategies for mediating between these two regimes of visibility. On the one hand, by employing an aesthetic that has been popularized by some of the most successful female black musicians of the age (from Beyoncé to Janelle Monáe) the Clermont twins are maximizing the potential of their images and videos to be 'surfaced' by algorithmic search engines. It is strategy of commercial visibility. But on the other hand, Afrofuturist aesthetics are employed to intervene into the role of race in structuring the visual field. As Browne has shown, black bodies have historically been rendered highly visible by surveillance systems while paradoxically being invisibilized politically, denied access to the categories of political personhood and the full status of the human. The Afrofuturist twin selfie enacts a double displacement of the liberal subject: presenting the twins as entangled both with each other's bodies and the digital networks of social media. They claim visibility in order to divert its attention away from dominant categories of the subject towards alternative configurations of bodies and images.

Conclusion: Deathly repetition

In photographs of twin faces, it is not sameness per se that makes them so fascinating but rather the complex interplay between sameness and difference. This dynamic is exposed by what is probably the best-known image of twins in the history of photography: 'Identical Twins, Roselle, New Jersey' taken by Diane Arbus in 1967. Arbus's photograph has had a subtly pervasive influence on visual representations of twins, being the inspiration behind the butchered twins in Stanley Kubrick's 1980 adaptation of the Stephen King novel *The Shining* and the motivation for countless parodies, including Sandro Miller's image of John Malkovich as both sisters. Both the Malkovich photographs – one in a series in which photographer Sandro Miller and the actor recreate some of the photographic iconic images of the twentieth century – and the image from *The Shining* emphasize the troubling qualities of the original Arbus photograph.

Like all photos of identical twins, when first contemplated, the image of the Roselle sisters impacts the viewer in various stages. The first thing you notice is the apparent symmetry. Both siblings, Cathleen and Colleen Wade, aged seven when the photograph was taken, are stood shoulder-to-shoulder facing the camera and wearing the same sombre corduroy dresses with white head bands over identical haircuts. The most immediately noticeable aspects of the framing of the image also emphasize this symmetry. The photograph is taken at waist level (head-height for the girls) using a Rolleiflex camera that creates a symmetrically square image. (Arbus's favoured Rolleiflex is endowed with a twin lens producing a parallel between the technology she used and the theme of duplication that she returned to throughout her career.) But the more you look at the photograph the more you notice the lack of symmetry. The line where the brick paving meets the whitewashed wall that provides the background of the photograph is at an angle, giving the photograph as a whole a feeling of being off-kilter. This imbalance draws the eye to the points of difference between the two girls. The tights they are wearing bear different patterns; their bone structures are also slightly different, producing subtly contrasting facial contours. Finally, Colleen on the right looks back at the camera with a limpid smile – a clarity of expression that is heightened by the fuller manner in which the light hits her face – while Cathleen seems a bit more sombre (in both literal and figurative senses) and wary of the photographer. It is this difference of expression that is emphasized by Malkovich who glares as Cathleen and beams as Colleen. As a result of this exaggeration, Miller's photograph is not just a parody but an interpretation that reduces the subtle play of difference in the Arbus image to contrasting comedy and tragedy masks.

The most persuasive interpretation of the photograph is provided by the US critic Carol Armstrong who sees the play of similarity and difference between the Roselle twins as key to Arbus's entire oeuvre. Several of the iconic images produced by Arbus are characterized by a near symmetric composition that is thrown off by the details. 'A husband and wife in the woods in a nudist camp, N.J.' (1963), for instance, shows a couple stood side-by-side facing the camera. Despite its square aspect ratio, the image is subtly asymmetrical. The man stands not fully face-on but at a slight angle while trees are clumped in an irregular manner behind them. 'What Arbus is interested in', Armstrong claims, is not symmetry, but 'the slight "flaw" that decentres symmetry'.[65] The way that the camera emphasizes the differences between the two twins in the 1967 photograph is highly significant. For Armstrong, this is the key that unites Arbus's theory of the photographic 'reproduction' of reality and the theme of biological reproduction that dominates her work through her two main interests in families (couples, parents and sibling relationships) and, to use the word most commonly associated with Arbus, 'freaks'. In Armstrong's words, 'Identical Twins, Roselle, New Jersey' 'quite literally shows how sameness mutates into difference by means of the flaw at both the levels of biology and of photography'.[66] Just as mutation is inherent to biological reproduction – nothing in nature is reproduced exactly; there is always difference, always change – photography does not produce exact copies of reality; rather, reality is changed by the fact of its visual reproduction and continues to mutate each time the image is reproduced and viewed in a different technological and cultural context.

In other words, through Arbus's framing, twinship precedes the simulacral logic of the image world. This construction of twinship as simulacral lies at the heart of the challenge to face recognition technology in the Face ID videos discussed throughout this chapter. In the Dolan Twins' video, the repetition of the sibling's faces is presented as foreshadowing the proliferating copies of their faces across social media platforms. And, as the twins are the product of the same split zygote, they are copies with no original. The videos, like the Arbus photograph, evoke the interplay between difference and repetition that is central to Deleuze's concept of the simulacrum. In his 1968 book *Difference and Repetition*, Deleuze constructs a concept of repetition not as the recurrence of a form or identity that remains unchanged through the process of iteration. Rather, repetition is described as a 'deathly' process that, in Colebrook's words, 'destroys sameness and unity' and 'abandons the self'.[67] The repetition set in motion by the twins' faces – a repetition that echoes through the logic of algorithmic image culture – undoes identity and unmoors the self. It is this simulacral process that the twin Face ID videos hold up as a challenge to the abstract machine of face recognition technology. While biometric systems attempt to reinforce the ties that bind the proliferating representations (both photographic images and the data shadows attached to them) to embodied individuals, images of twins circulating through social media networks loosen those bonds, opening them up to a logic of virtual becoming.

The blurred distinction between original and copy in the simulacral status of twinship in social media stardom underscores the disruptive quality of the biometric failure of twin faces. Central to Hansen's argument that aesthetic experiments with the Digital Facial Image produce possibilities of 'new affective relations' between humans and computer systems is the concept of 'affectivity' or 'affection' that he borrows from Gilbert Simondon (1989) and Deleuze (1986). Both describe a form of sensation that mediates between the individual body's habituated somatic responses and a 'preindividual reality' that, in Simondon's words, is 'to a certain extent heterogeneous in relation to the individualized reality [...] indicating to the individualized being that it is not a complete and closed set of reality'.[68] In new media artworks such as *Dream of Beauty 2.0*, Hansen argues, 'affectivity becomes the very medium of the interface with the image'.[69] As mediator, this affectivity 'actualizes the potential of the image at the same time as it virtualizes the body: the crucial element is neither image nor body alone, but the dynamic interaction between the two'.[70] The affection produced by the aesthetic encounter with the DFI blurs the distinction between human subject and computational object. The affectivity of the disruption introduced into Apple's biometric interface by twin faces has a similar effect in that it undermines binary distinctions between human and machine that are fundamental to face recognition systems more generally. As Zara Dinnen and Sam McBean put it, face recognition 'relies on both the certainty of human-machine binaries (that there is an autonomous face to be recognized by an autonomous program) and on the premise these autonomies can be overturned by the machine (that the face can be "understood" by the software)'.[71] Twin faces disrupt these binaries by introducing glitches into the interface and by presenting the simulacral logic of algorithmic image culture as 'naturally' pre-existing within the twins themselves. Twinship, as a trope so frequently used to interrogate binary thinking of various kinds, disrupts the oppositions between human and machine that are central to the neoliberal logic applied to digital culture and confront us with the conceptual challenge of entanglement.

Notes

1   Other consumer tech reviewers employed similar strategies with different results. *Tech Insider*, for instance, tested the system using thirty-six-year-old twin brothers and concluded: 'You can't fool the iPhone X.' 'We Put the iPhone X's Face ID to the Test with Twins', *YouTube*, 31 October 2017, https://www.youtube.com/watch?v=rFoSPZBHsmE. But the majority of 'Twin Tests' posted on YouTube found that identical twin faces could unlock each other's phones. When one twin unlocks his brother's phone on a video posted by Mashable, his sibling responds: 'Don't steal my data, no! [...] The iPhone X is not twin compatible.' 'Is the iPhone X's Facial Recognition Twin Compatible?', *YouTube*, 31 October 2017, https://www.youtube.com/watch?v=e8-yupM-6Oc.
2   Lisa Zunshine, *Strange Concepts and the Stories They Make Possible* (Baltimore: The Johns Hopkins University Press, 2008), 35.

3   Lila Lee-Morrison, *Portraits of Automated Facial Recognition: On Machinic Ways of Seeing the Face* (Bielefeld: Transcript Verlag, 2019), 45.
4   Kelly Gates, *Our Biometric Future: Facial Recognition Technology and the Culture of Surveillance* (New York and London: New York University Press, 2011), 14.
5   William Viney, 'Getting the Measure of Twins', in *The Edinburgh Companion to the Critical Medical Humanities*, ed. Anne Whitehead, Angela Woods, Sarah Atkinson, Jane Macnaughton and Jennifer Richards (Edinburgh: Edinburgh University Press, 2016), 107.
6   Patrick Grother, Mei Ngan and Kayee Hanaoka, 'Ongoing Face Recognition Vendor Test Part 2: Identification', National Institute of Standards and Technology, 2018, https://doi.org/10.6028/NIST.IR.8238.
7   Grother, Ngan and Hanaoka, 'Ongoing Face Recognition Vendor Text Part 2: Identification', 29.
8   Kevin W. Bowyer and Patrick J. Flynn, 'Biometric Identification of Identical Twins: A Survey', Department of Computer Science and Engineering, University of Notre Dame, 2016, https://www3.nd.edu/~kwb/Bowyer_Flynn_BTAS_2016.pdf.
9   Jussi Parikka, *Digital Contagions: A Media Archaeology of Computer Viruses: Second Edition* (New York: Peter Lang, 2016), xxxiii.
10 Parikka, *Digital Contagions*, xxxiii.
11 Parikka, *Digital Contagions*, xxxiii.
12 Michel Foucault, *Discipline and Punish: The Birth of the Prison*, trans. Alan Sheridan (London: Penguin, 1991 [1975]), 201.
13 Susie Khamis, Lawrence Ang and Raymond Welling, 'Self-Branding, "Micro-Celebrity" and the Rise of Social Media Influencers', *Celebrity Studies* 8:2 (2017): 191–208.
14 Daniela Agostinho, 'The Optical Unconscious of Big Data: Datafication of Vision and Care for Unknown Futures', *Big Data & Society* (January–June 2019): 2.
15 Agostinho, 'The Optical Unconscious of Big Data', 2.
16 Agostinho, 'The Optical Unconscious of Big Data', 4.
17 Tama Leaver, Tim Highfield and Crystal Abidin, *Instagram: Visual Social Media Cultures* (Cambridge: Polity, 2020), n/a.
18 Alexis C. Madrigal, 'No, You Don't Really Look Like That', *The Atlantic*, 16 December 2018, https://www.theatlantic.com/technology/archive/2018/12/your-iphone-selfies-dont-look-like-your-face/578352/?utm_source=facebook&utm_campaign=the-atlantic&utm_content=edit-promo&utm_term=2018-12-18T20%3A54%3A27&utm_medium=social.
19 Madrigal, 'No, You Don't Really Look Like That.'
20 Daniel Rubinstein and Katrina Sluis, 'The Digital Image in Photographic Culture: Algorithmic Photography and the Crisis of Representation', in *The Photographic Image in Digital Culture: Second Edition*, ed. Martin Lister (London and New York: Routledge, 2013), 35.

21 Rubinstein and Sluis, 'The Digital Image in Photographic Culture', 35.
22 James Ash, Ben Anderson, Rachel Gordon, and Paul Langley, 'Digital Interface Design and Power: Friction, Threshold, Transition', *Environment and Planning D: Society and Space* 36:6 (2018): 1139.
23 Sarah Kember, 'Face Recognition and the Emergence of Smart Photography', *Journal of Visual Culture* 13:2 (2014): 193.
24 Wendy Hui Kyong Chun, 'Queering Homophily', in Clemens Apprich, Wendy Hui Kyong Chun, Florian Cramer, and Hito Steyerl, *Pattern Discrimination* (Minneapolis and London: University of Minnesota Press, 2018), 75.
25 Chun, 'Queering Homophily', 66.
26 In a tweet critiquing the marketing discourse behind the machine learning biometrics firm Faception: Facial Personality Analytics, Ben Snyder comments: 'Today in shitty machine learning start-ups, this company claims to predict IQ, personality, and violent tendencies by applying deep learning to facial features and bone structure. That's phrenology. You just made the ML equivalent of a racist uncle.' (20 November 2018)
27 Simone Browne, 'Digital Epidermalization: Race, Identity and Biometrics', *Critical Sociology* 36:1 (2009): 132.
28 Paul Gilroy, *Against Race: Imagining Political Culture beyond the Color Line* (Cambridge, MA: Harvard University Press, 2000), 37.
29 Lee-Morrison, *Portraits of Automated Facial Recognition*, 19.
30 Claire Colebrook, *Death of the PostHuman: Essays on Extinction, Vol. 1* (Ann Arbor: Open Humanities Press, 2014), 150.
31 Colebrook, *Death of the PostHuman*, 140.
32 Colebrook, *Death of the PostHuman*, 148.
33 Colebrook, *Death of the PostHuman*, 150.
34 Colebrook, *Death of the PostHuman*, 150.
35 Jay Owens, 'The Age of Post-Authenticity and the Ironic Truths of Meme Culture', *Medium*, 11 April 2018, https://medium.com/s/story/post-authenticity-and-the-real-truths-of-meme-culture-f98b24d645a0.
36 In his study of the use of twins in advertising images, Hillel Schwartz argues that the dominant photographic convention of displaying identical twins hip-to-hip presents them as a 'synecdoche of mass production.' Identical twins are presented as nature's counterpart for the standardized industrial mass production techniques developed by Henry Ford. Not coincidentally, Ford Motor Cars itself once used images of twins to promote its vehicles. Schwartz, *The Culture of the Copy*, 39.
37 Peter Zelewski, *Twins* (London: Hoxton Mini Press, 2018).
38 Audrey Niffenegger, *Her Fearful Symmetry* (London: Jonathan Cape, 2009).
39 The use of twins to indicate the limits of human vision has a longer history than that of photography. In his discussion of twins and doubles, Peter Sloterdijk

makes a connection between ancient spiritual traditions of imagining the soul as something external to the individual – either a 'divine milieu' encasing the individual or a guardian spirit – and the appearance of twin angels in ancient Christianity. Legends about Saint Anthony collected in the *Apophthegmata Patrum* or *Sayings of the Desert Fathers* (a collection of over a 1000 brief stories and sayings compiled in the fifth and sixth centuries) provide a blueprint for the use of the double or twin to represent a general tendency towards personalized angels: the idea that all of God's children are accompanied by their very own angel. In this tradition, twins were used to represent the spiritual doubles of earthly beings. Peter Sloterdijk, *Spheres Volume I: Bubbles: Microspherology*, trans. Wieland Hoban (Los Angeles, CA: Semiotext(e), 2011), 425–6.
40 Marianne Hirsch, *Family Frames: Photography, Narrative and Postmemory* (Cambridge, MA and London: Harvard University Press, 1997), 97.
41 Zach Blas, 'Escaping the Face: Biometric Facial Recognition and the Facial Weaponization Suite', *Media-N* 9:2 (2013): n/a.
42 Blas, 'Escaping the Face'.
43 Blas, 'Escaping the Face'.
44 Jenny Edkins, *Face Politics* (London: Routledge, 2015), 3.
45 Edkins, *Face Politics*, 3.
46 Edkins, *Face Politics*, 7.
47 Zygmunt Bauman, *Liquid Modernity* (London: Polity, 2000), 51–2.
48 Stephen Marche, 'Modern Family', *Times Literary Supplement*, 22 August 2018.
49 Aaron Hess, 'The Selfie Assemblage', *International Journal of Communication* 9 (2015): 1630.
50 Hess, 'The Selfie Assemblage', 1631.
51 Paul Frosh, 'The Gestural Image: The Selfie, Photography Theory and Kinaesthetic Sociability', *International Journal of Communication* 9 (2015): 1608.
52 Frosh, 'The Gestural Image', 1611.
53 Frosh, 'The Gestural Image', 1612.
54 Frosh, 'The Gestural Image', 1609.
55 Elizabeth Grosz, 'Intolerable Ambiguity: Freaks as/ at the Limit', in *Freakery: Cultural Spectacles of the Extraordinary Body*, ed. Rosemarie Garland Thomson (New York and London: New York University Press, 1996), 65.
56 Grosz, 'Intolerable Ambiguity', 65.
57 Gilles Deleuze and Félix Guattari, *A Thousand Plateaus: Capitalism and Schizophrenia*, trans. Brian Massumi (London: Continuum, 2004 [1980]), 187.
58 Deleuze and Guattari, *A Thousand Plateaus*, 189–90.
59 Gilles Deleuze, *Cinema I: The Movement-Image*, trans. Hugh Tomlinson and Barbara Habberjam (Minneapolis: University of Minnesota Press, 1986), 96.

60 Mark Hansen, 'Affect as Medium, or the "Digital-Facial-Image"', *Journal of Visual Culture* 2:2 (2003): 206.
61 Hansen, 'Affect as Medium', 209.
62 Hansen, 'Affect as Medium', 208 and 209.
63 Celia Lury, 'Brand as Assemblage', *Journal of Cultural Economy* 2: 1–2 (2009): 67.
64 Daniela Agostinho, 'Chroma Key Dreams: Algorithmic Visibility, Fleshy Images and Scenes of Recognition', *Philosophy of Photography* 9:2 (2018): 132.
65 Carol Armstrong, 'Biology, Destiny, Photography: Difference According to Diane Arbus', *October 66* (1993): 33.
66 Armstrong, 'Biology, Destiny, Photography', 34.
67 Gilles Deleuze, *Difference and Repetition*, trans. Paul Patton (London: Bloomsbury, 2014), 135.
68 Gilbert Simondon, *L'Individuation Psychique et Collective* (Paris: Aubier, 1989), 108.
69 Hansen, 'Affect as Medium', 208.
70 Hansen, 'Affect as Medium', 208.
71 Zara Dinnen and Sam McBean, 'The Face as Technology', *New Formations* 93 (2017): 125.

Conclusion: From digital twins to glitch twins

In April 2021, a project called the Earth Archive Initiative announced a plan to create a digital twin of the entire surface of the Earth. Starting with the Amazon basin, scientists from around the world will employ LiDAR (Light Detection and Ranging) technology, which uses laser beams to measure distances in space, to create an exact 3D replica of the Earth's surface. The aim is to create a record for future geologists to precisely assess the degree and nature of erosion and tectonic shifts. But it is also intended as a tool in the fight against ecological crisis. A digital twin of the Earth's surface could be the baseline for a more complex virtual replica of Earth's ecosystems and this could be used to model potential future climate change scenarios and mitigate against disaster. The fact that the Earth Archive Initiative mobilizes the metaphor of the digital twin in their fund-raising campaign is revealing of how much currency the concept has in a range of industries from engineering to healthcare.

In the same year, Rolls-Royce publicized its pioneering use of digital twin technology in its aircraft engines. Their new generation of engines contains sensors that capture hundreds of data points every second, which are analysed in real time using machine learning and AI tools. According to Chief Information Officer Stuart Hughes, this allows Rolls-Royce to improve the sustainability of its in-field engines, allowing them to provide up-to-date guidance to pilots. The digital twin system also enables 'predictive maintenance' through a 'personal approach to understanding how each engine is flown, the context it's flown within [and] the environmental conditions'.[1] Digital twins help to 'optimise' performance through a continual feedback loop between physical systems and the analysis of data produced by these systems. This strategy is also set to revolutionize healthcare. General Electric's Digital CEO Bill Ruh has predicted that in the not-too-distant future everybody will be given a digital twin at birth. This will 'take data' from wearables and sensors embedded within the body which will be used to predict and prevent the onset of disease.[2] As well as providing a 'guide to health and personal maintenance', as Mark Minevich puts it, personalized digital twins might also provide direction for 'performance enhancements'.[3]

The promise of digital twins is the ability to see into the future. And if you can see into the future you can control it. As Mark Girolami puts it, the idea is that creating an 'abstracted representation of a physical or natural process or system' will allow us to 'make predictions about that reality; to make forecasts and to control it'.[4] This use of twins as a metaphor returns, once again, to what Hillel Schwartz describes as the 'oracular power of twins'.[5] Just as, among the Lele of the Congo, for example, parents of twins become diviners, genetics research uses the twin method to foresee patterns of human development. When the twin method rose to prominence in the 1920s, 'twins became Controls, the ongoing life and body of one held as duplicate register against the other, or the lives and bodies of fraternals set against identicals'.[6] The concept of digital twins is also indicative of the enduring appeal of twins as a guiding metaphor of cybernetics. The metaphor of the digital twin, as it is most often used in engineering, highlights the danger of cybernetics as a totalizing mode of thought. As Yuk Hui puts it, 'cybernetic thinking remains a thinking of totalization, since it aims to absorb the other into itself, like Hegelian logic, which sees polarity not as oppositional but rather as a motivation towards synthesized identity'.[7] The digital twin is a figure of technological control in which the oracular nature of twinship is conscripted into a technological rationality. The future is fully predictable and controllable.

In opposition to this techno-rationalist use of twins, in this book I have traced an alternative discourse that aligns twinship with the dynamics of glitching. Reporting on discussions at the 2010 GLI.TC/H conference, Tom McCormack points to a founding moment when conference attendees first became interested in glitch aesthetics: 'the moments in their childhood when their video game systems would glitch, offer[ed] up a glimpse at the hidden world of data structuring this new mediated experience'.[8] In constituting moments of failure, glitches reveal the rules governing systems and the infrastructures that are usually hidden from view by commercial packaging. Rather than transparent windows onto the agency of the past through inherited generic traits, glitch twins are a form of weird media that introduce blockages into dynamics of mediation. In the process, they draw attention to the logics governing these dynamics. The weird twins discussed in Chapter 2 introduce blockages into the mediating role of genetics research, while the social media twin stars examined in Chapter 6 perform a failure in the biometric systems that interface computer systems with human users. These moments of failure open a space of critique between the twin demands of individuality and dividuality that regulate neoliberal cybernetic control systems. As a result, they reveal alternate ways of conceptualizing these systems and produce new socio-technological configurations. Curt Cloninger points out that 'the glitch foregrounds and problematizes [the] myth of pure transcendental data, of pure and perfect signal. The glitch is a perpetual reminder of the immanent, real-time embodiment of executed code.'[9] While twin studies reduce its twins to scientific instruments, channels of data about underlying genetic codes, glitch twins insist on the contingencies of embodiment and the continual feedback loops between bodies and environments.

Rather than predictable and controllable, the future heralded by glitch twins is open, characterized by weird loops that are unpredictable and uncontrollable. The twins in the works of speculative black futurism discussed in Chapter 4, such as the French dance duo Les Twins, loop together traumatic generational memories of transatlantic slavery with a technological future that displaces the anti-Blackness encoded within the dominant narratives of modernity in the twentieth century. The melancholy twins of Chapter 2 position the past as a blockage within the present. The narrators of *Half Life*, the mutant twin progeny of ecological crisis discussed in Chapter 3, enact a moebius strip-like timeline that undermines modern concepts of linear temporality, while modelling a distributed, non-anthropocentric mode of subjectivity. Far from predictable, these futures are unstable and flicker in and out of view. Furthermore, by emerging in the moments of transition between scales, glitch twins provide a conceptual tool for the 'scale-shifting' required by critical responses to the climate crisis, a lens through which to envision connections between the human, the micro-scale of cellular structures and the macro-scale of planetary processes. But, while they become a lens for thinking across scales, glitch twins themselves confound the logic of the smooth scalability envisaged by industrial capitalism.

One thing that the narratives and images of glitch twins examined in this book do have in common with the digital twins of engineering optimization is their appeal to the language of myth. In Chapter 1, I argued that twins have been a central figure in the 'mythotechnesis' of cybernetics, a term that David Burrows and Simon O'Sullivan use to describe 'practices which instantiate or perform [...] different human-machine relations'.[10] Grey Walter resorted to the language of twins when describing the interactions between his robots Elmer and Elsie. The glitching dance that they perform when interacting with each other was held as an instance of unprogrammed emergent behaviour and an example of the parallels between the processual ontologies of machinic and organic organisms.

Within the cybernetic narratives discussed throughout the chapter, twins are not the embodiment of deathly repetition of closed systems but of a dynamic and productive opening out that tests the limits of control systems. In their use of twins, these scientists and writers foreshadow the way that genetics studies of the 1990s harnessed the associations between twins and myth to publicize their controversial research findings. But the appeal to myth draws attention to a tension that runs through the texts and images studied in this book. While one of the main functions of myth is to naturalize historically contingent narratives about the foundations of collective ethnic or national identities, glitches serve the anti-mythical function of revealing the contingencies and materialities of data infrastructures.

This, in turn, points to an apparent series of contradictions in the deployment of twins within the discourses traced through this book: the use of twins, often taken as the embodiment of dualisms of various kinds, to overcome binary thought; the attempt to articulate non-modern and non-anthropocentric modes of being through Amerindian beliefs about twins while repeating a colonizing logic that strips indigenous thought of its historical and embodied contexts. This pattern of contradiction goes to the heart of twinning as a process. Florence Chiew and Alison Barnwell propose the figure of twins as a model for what they describe as 'methodological intimacy': a way to think through the entanglements between analytical distance and affective proximity that are present in any critical project. Since twins 'embody at once a sense of dualism and entanglement, unsettling easy divisions between nature and nurture, self-identity and difference', they provide a blueprint for a mode of criticism that simultaneously suspends and reaffirms divides between the subject and object of thought. So, while the representations of twins explored in this book are deployed to define the contours and test the limits of the systems that define what it means to be human, to model emerging modes of intrapersonal connection and new configurations of humans and technology, they also challenge us to rethink the methodologies of critical thought itself.

Notes

1 Quoted in Derek du Preez, 'How Rolls-Royce Is Improving Engine Sustainability with Real-time Data and Digital Twins,' *Diginomica*, 5 April 2021, https://diginomica.com/how-rolls-royce-improving-engine-sustainability-real-time-data-and-digital-twins.

2   Bill Ruh quoted in Kim Bellard, 'Twins for Everyone,' *Medium*, 15 June 2021, https://kimbellard.medium.com/twins-for-everyone-9837dc7620f.
3   Mark Minevich, 'Digital Twin Technology May Save Lives and Solve the Biggest Challenges Facing Humanity', *The Hill*, 22 May 2021, https://thehill.com/opinion/technology/554881-digital-twin-technology-may-save-lives-and-solve-the-biggest-challenges.
4   Mark Girolami, 'Digital Twins: The Sense and Statistics,' Keynote Address, Royal Statistical Society, 29 October 2020, https://www.youtube.com/watch?v=cWIJNXQn8LI.
5   Hillel Schwartz, *The Culture of the Copy: Striking Likenesses, Unseasonable Facsimiles* (New York: Zone Books, 1996), 34.
6   Schwartz, *The Culture of the Copy*, 34.
7   Yuk Hui, 'Machine and Ecology', *Angelaki* 25:4 (2020), 63.
8   Tom McCormack, 'Tom McCormack, "Code Eroded: At GLI.TC/H 2010, RHIZOME,"' in *GLI.TC/H 20111: READER[ROR]*, ed. Nick Briz, Evan Meaney, Rosa Menkman, William Robertson, Jon Satrom, Jessica Westbrook (Chicago: Unsorted Books, 2011), 18.
9   Curt Cloninger, 'GltchLnguistx: The Machine in the Ghost / Static Trapped in Mouths,' in *GLI.TC/H 20111: READER[ROR]*, 24.
10 David Burrows and Simon O'Sullivan, *Fictioning: The Myth-Functions of Contemporary Art and Philosophy* (Edinburgh: Edinburgh University Press, 2017), 342.

Works Cited

12 Monkeys (1995), [Film] Dir. Terry Gilliam, USA: Atlas Entertainment.
Abels, Birgit. (2019), 'A Poetics of Dwelling with Music and Dance: Le hip hop as Homing Practice', *The World of Music*, 8 (1): 49–64.
Achebe, Chinua. (2010 [1958]), *Things Fall Apart*, London: Penguin.
Agostinho, Daniela. (2019), 'The Optical Unconscious of Big Data: Datafication of Vision and Care for Unknown Futures', *Big Data & Society*, (January–June): 1–10.
Agostinho, Daniela. (2018), 'Chroma Key Dreams: Algorithmic Visibility, Fleshy Images and Scenes of Recognition', *Philosophy of Photography*, 9 (2): 131–55.
Aljana Moons (2015), [Film] Dir. Alexis Peskine, France: independent.
Alaimo, Stacy. (2016), *Exposed: Environmental Politics & Pleasure in Posthuman Times*, Minneapolis: University of Minnesota Press.
Aldiss, Brian. (2012), *Brothers of the Head*, London: The Friday Project.
Alexander, Elizabeth. (1994), '"Can You Be Black and Look at This?" Reading the Rodney King Video(s)', *Public Culture*, 7 (1): 77–94.
Anderson, Reynaldo and Charles E. Jones. (2016), 'Introduction: The Rise of Astro-Blackness', in Reynaldo Anderson and Charles E. Jones (eds), *Afrofuturism 2.0: The Rise of Astro-Blackness*, vii–xvii, New York: Lexington Books.
Ansell-Pearson, Keith. (1999), *Germinal Life: The Difference and Repetition of Deleuze*, London and New York: Routledge.
Armstrong, Carol. (1993), 'Biology, Destiny, Photography: Difference According to Diane Arbus', *October*, 66: 28–54.
Ash, James, Ben Anderson, Rachel Gordon, and Paul Langley. (2018), 'Digital Interface Design and Power: Friction, Threshold, Transition', *Environment and Planning D: Society and Space*, 36 (6): 1136–53.
Ayorech, Ziada, Sophie von Stumm, Claire Haworth, Oliver Davis and Robert Plomin. (2017), 'Personalized Media: A Genetically Informative Investigation of Individual Differences in Online Media Use', *PLoS ONE*, 12 (1): https://doi.org/10.1371/journal.pone.0168895 (accessed 15 November 2018).
Azuma, Hiroki. (2009), *Otaku: Japan's Database Animals*, Minneapolis: University of Minnesota Press.
Bargen, Doris G. (1991), 'Twin Blossoms on a Single Branch: The Cycle of Retribution in *Onnamen*', *Monumenta Nipponica*, 46 (2): 147–71.
Barrett, William F., et al. (1882–3), 'Report of the Literary Committee', *Proceedings of the Society for Psychical Research*, 1: 116–55.
Bastian, Misty L. (2001), '"The Demon Superstition": Abominable Twins and Mission Culture in Onitsha History', *Ethnology*, 40 (1): 13–27.

Baucom, Ian. (2005), *Specters of the Atlantic: Finance Capital, Slavery, and the Philosophy of History*, Durham, NC: Duke University Press.

Bauman, Zygmunt. (2000), *Liquid Modernity*, London: Polity.

Behrend, Heiki and Martin Zillinger. (2015), 'Introduction: Trance Mediums and New Media', in Heiki Behrend, Anja Dreschke, and Martin Zillinger (eds), *Trance Mediums and New Media: Spirit Possession in the Age of Technical Reproduction*, 1–24, New York: Fordham Scholarship Online.

Bellard, Kim. (2021), 'Twins for Everyone', *Medium*, 15 June. Available online: https://kimbellard.medium.com/twins-for-everyone-9837dc7620f (accessed 29 June 2021).

Bhabha, Homi. (1994), *The Location of Culture*, London and New York: Routledge.

The Bight of the Twin (2016), [Film] Dir. Hazel Hill McCarthy III, USA.

Blas, Zach. (2013), 'Escaping the Face: Biometric Facial Recognition and the Facial Weaponization Suite', *Media-N* 9 (2).

Boklage, Charles E. (2010), *How New Humans Are Made: Cells and Embryos, Twins and Chimeras, Left and Right, Mind/Self/Soul, Sex, and Schizophrenia*, Singapore: World Scientific Publishing.

Bollmer, Grant. (2016), *Inhuman Networks: Social Media and the Archaeology of Connection*, London: Bloomsbury.

Bould, Mark. (2021), *The Anthropocene Unconscious: Climate Catastrophe in Contemporary Culture*, London: Verso.

Bowyer, Kevin W. and Patrick J. Flynn. (2016), 'Biometric Identification of Identical Twins: A Survey', Department of Computer Science and Engineering, University of Notre Dame. Available online: https://www3.nd.edu/~kwb/Bowyer_Flynn_BTAS_2016.pdf (accessed 15 November 2018).

Briz, Nick. (2009), 'Thoughts on Glitch[Art]v2.0'. Available online: http://nickbriz.com/thoughtsonglitchart/ (accessed 29 June 2021).

Broderick, Damien. (2018), *Psience Fiction: The Paranormal in Science Fiction Literature*, Jefferson, NC: McFarland & Company.

Brooklyn and Bailey. (2017), 'Twin Which Twin: Guess the Baby Pictures?', *YouTube*, 22 February, https://www.youtube.com/watch?v=ppi-c81j7jM (accessed 7 July 2021).

Browne, Simone. (2009), 'Digital Epidermalization: Race, Identity and Biometrics', *Critical Sociology*, 36 (1): 131–50.

Burrows, David and Simon O'Sullivan. (2017), *Fictioning: The Myth-Functions of Contemporary Art and Philosophy*, Edinburgh: Edinburgh University Press.

Burton, James. (2015), 'From Exegesis to Ecology', in Alexander Dunst and Stefan Schlensag (eds), *The World According to Philip K. Dick*, 209–27, New York: Palgrave Macmillan.

Butler, Mark. (2006), *Unlocking the Groove: Rhythm, Meter, and Musical Design in Electronic Dance Music*, Bloomington: Indiana University Press.

Cassegard, Carl. (1999), 'Shock and Modernity in Walter Benjamin and Kawabata Yasunari', *Japanese Studies*, 19 (3): 237–51.

Cannon, Kristopher L. (2014), 'Ec-statically Queer Images: Queering the Photographic through Fetal Photography', *Photography and Culture*, 7 (3): 269–83.
Chandler, David and Julian Reid. (2020), 'Becoming Indigenous: The "speculative turn" in Anthropology and the (Re)colonisation of Indigeneity', *Postcolonial Studies*, 23 (4): 485–504.
Chaudhuri, Una. 'Enduring Performance', Keynote Address at 'Art in the Anthropocene' conference, Trinity College, Dublin, 8 June 2019.
Cheney-Lippold, John. (2011), 'A New Algorithmic Identity: Soft Biopolitics and the Modulation of Control', *Theory, Culture & Society*, 28 (6): 164–81.
Chiew, Florence and Alison Barnwell. (2019), 'Methodological Intimacies and the Figure of the Twins', *The Sociological Review Monographs*, 67 (2): 467–80.
Chun, Wendy Hui Kyong. (2018), 'Queering Homophily', in Clemens Apprich, Wendy Hui Kyong Chun, Florian Cramer and Hito Steyerl (eds), *Pattern Discrimination*, 59–97, Minneapolis and London: University of Minnesota Press.
Clarke, Bruce. (2009), 'Heinz von Foerster's Demons: The Emergence of Second-Order Systems Theory', in Clarke and Mark B. N. Hansen (eds), *Emergence and Embodiment: New Essays on Second-Order Systems Theory*, 34–61, Durham and London: Duke University Press.
Clarke, Bruce and Mark B. N. Hansen. (2009), 'Introduction: Neocybernetic Emergence', in Bruce Clarke and Mark B. N. Hansen (eds), *Emergence and Embodiment: New Essays on Second-Order Systems Theory*, 1–25, Durham and London: Duke University Press.
Clark, Timothy. (2012), 'Scale', in Tom Cohen (ed.), *Telemorphosis: Theory in the Era of Climate Change, Vol. 1*, 148–66, Ann Arbor: Open Humanities Press. Available online: http://www.openhumanitiespress.org/books/titles/telemorphosis/ (accessed 29 June 2021).
Cloninger, Curt. (2011), 'GltchLnguistx: The Machine in the Ghost/Static Trapped in Mouths', in Nick Briz, Evan Meaney, Rosa Menkman, William Robertson, Jon Satrom and Jessica Westbrook (eds), *GLI.TC/H 20111: READER[ROR]*, 23–41, Chicago: Unsorted Books.
Colebrook, Claire. (2014), *Death of the PostHuman: Essays on Extinction, Vol. 1*. Ann Arbor: Open Humanities Press.
Colwell, C. (1996), 'Discipline and Control: Butler and Deleuze on Individuality and Dividuality', *Philosophy Today*, 40 (1): 211–16.
Cooper, Brenda. (2008), 'Diaspora, Gender and Identity: Twinning in Three Diasporic Novels', *English Academy Review*, 25 (1): 51–65.
Cornyetz, Nina. (2009), 'Fascist Aesthetics and the Politics of Representation in Kawabata Yasunari', in Alan Tansman (ed.), *The Culture of Japanese Fascism*, 321–54, Durham, NC: Duke University Press.
Danowski, Déborah and Eduardo Viveiros de Castro. (2017), *The Ends of the World*, trans. Rodrigo Nunes, Cambridge: Polity Press.
Davis, Erik. (2015), 'High Weirdness: Visionary Experience in the Seventies Counterculture', Diss., Rice University, Houston, Texas.

Dead Ringers (1988), [Film] Dir. David Cronenberg, USA: Morgan Creek Entertainment.
Deleuze, Gilles. (1986), *Cinema I: The Movement-Image*, trans. Hugh Tomlinson and Barbara Habberjam. Minneapolis: University of Minnesota Press.
Deleuze, Gilles. (1995), 'Postscript on the Society of Control', in *Negotiations: 1972–1990*, trans. Martin Joughlin, 177–82, New York: Columbia University Press.
Deleuze, Gilles. (2014), *Difference and Repetition*, trans. Paul Patton. London: Bloomsbury.
Deleuze, Gilles and Félix Guattari. (2004), *A Thousand Plateaus: Capitalism and Schizophrenia*, trans. Brian Massumi, London: Continuum.
Dick, Philip K. (2001 [1981]), *Valis*, London: Gollancz.
Dick, Philip K. (2007 [1968]), *Do Androids Dream of Electric Sheep?*, London: Gollancz.
Dick, Philip K., Pamela Jackson and Jonathan Lethem (eds) (2011), *The Exegesis of Philip K. Dick*, London: Gollancz.
Dillon, Karen. (2018), *The Spectacle of Twins in American Literature and Popular Culture*, Jefferson, NC: McFarland & Company, Inc.
Dinnen, Zara and Sam McBean. (2017), 'The Face as Technology', *New Formations*, 93: 122–36.
Dolan Twins. (2017), 'Twins Vs. iPhone X Face ID', *YouTube*, 4 November. Available online: https://www.youtube.com/watch?v=GFtOaupYxq4 (accessed 7 July 2021).
The Double (2013), [Film] Dir. Richard Ayoade, UK: Alcove Entertainment.
DuBois, W.E.B. (1994), *The Souls of Black Folk*, Mineola, NY: Dover Publications.
Edkins, Jenny. (2015), *Face Politics*, London: Routledge.
Eglash, Ron. (1999), *African Fractals: Modern Computing and Indigenous Design*, New Brunswick and London: Rutgers University Press.
Emecheta, Buchi. (1994), *Kehinde*, Oxford: Heinemann.
Enchi, Fumiko. (1983 [1958]), *Masks*, trans. Juliet Winters Carpenter, New York: Vintage Books.
Enemy (2013), [Film] Dir. Denis Villeneuve, France: Pathé International.
Engelstein, Stefani. (2017), *Sibling Action: The Genealogical Structure of Modernity*, New York: Colombia University Press.
Eshun, Kodwo. (2003), 'Further Considerations on Afrofuturism', *The New Centennial Review*, 3 (2): 287–302.
Espinoza, Tania. (2013), 'The Technical Object of Psychoanalysis', in Christina Howells and Gerald Moore (eds), *Stiegler and Technics*, 151–64, Edinburgh: Edinburgh University Press.
Fisher, Mark. (2016), *The Weird and the Eerie*, London: Repeater Books.
Follet, Ken. (1996), *The Third Twin*, London: Pan Macmillan.
The Forest (2016), [Film] Dir. Jason Zada, USA: DragonCove Studios.
Foucault, Michel. (1991 [1975]), *Discipline and Punish: The Birth of the Prison*, trans. Alan Sheridan, London: Penguin.
Foucault, Michel. (2003 [1976]), *"Society Must Be Defended": Lectures at the Collège de France, 1975–76*, trans. David Macey, New York: Picador.

Frazer, James George. (1993 [1890]), *The Golden Bough: A Study in Magic and Religion*, London: Wordsworth Reference.
Free 6lack (2016), [Album] Artist 6BLACK, USA: LVRN and Interscope.
Freud, Sigmund. (1986 [1919]), 'The uncanny', in Anna Freud (ed.), *The Essentials of Psychoanalysis*, trans. James Strachey, 350–82, London: Pelican.
Freud, Sigmund. (2001 [1955]), 'Mourning and Melancholia (1917)', in James Strachey (ed.), *The Standard Edition of the Complete Psychological Works of Sigmund Freud (Vol. XVII)*, 243–58, London: Hogarth Press.
Freud, Sigmund. (2001 [1957]), 'Beyond The Pleasure Principle (1920)', in James Strachey (ed.), *The Standard Edition of the Complete Psychological Works of Sigmund Freud (Vol. XIV)*, 7–66, London: Hogarth Press.
Freud, Sigmund. (1961 [1923]), 'The Ego and the Id', in James Strachey (ed.), *The Standard Edition of the Complete Psychological Works of Sigmund Freud (Vol. XIX)*, 3–66, London: Hogarth Press.
Frosh, Paul. (2015), 'The Gestural Image: The Selfie, Photography Theory, and Kinesthetic Sociability', *International Journal of Communication*, 9: 1607–26.
Frost, Samantha. (2018), 'Ten Theses on the Subject of Biology and Politics: Conceptual, Methodological, and Biopolitical Considerations', in Maurizio Meloni, John Cromby, Des Fitzgerald and Stephanie Lloyd (eds), *The Palgrave Handbook of Biology and Society*, 897–23, New York: Palgrave.
Galison, Peter (1994), 'The Ontology of the Enemy: Norbert Wienr and the Cybernetic Vision', *Critical Inquiry*, 21 (1): 228–66.
Galton, Francis. (1885), 'The History of Twins as a Criterion of Nature and Nurture', *Fraser's Magazine*, 12: 566–76.
Gates, Kelly A. (2011), *Our Biometric Future: Facial Recognition Technology and the Culture of Surveillance*, New York and London: New York University Press.
Gilroy, Paul. (2000), *Against Race: Imagining Political Culture beyond the Color Line*. Cambridge, MA: Harvard University Press.
Girolami, Mark. (2020), 'Digital Twins: The Sense and Statistics', Keynote Address, Royal Statistical Society, 29 October. Available online: https://www.youtube.com/watch?v=cWIJNXQn8LI (accessed 29 June 2021).
Gomel, Elana and Stephen Weninger. (2003), 'Cronenberg, Greenaway and the Ideologies of Twinship', *Body & Society*, 9 (3): 19–35.
Greer, Amanda. (2017), 'Absence, Play, and the Antidetective Story: Shelley Jackson's *Half Life*', *Critique: Studies in Contemporary Fiction*, 59 (2): 168–79.
Grosz, Elizabeth. (1996), 'Intolerable Ambiguity: Freaks as/ at the Limit', in Rosemarie Garland Thomson (ed.), *Freakery: Cultural Spectacles of the Extraordinary Body*, 55–66, New York and London: New York University Press.
Grother, Patrick, Mei Ngan and Kayee Hanaoka. (2018), 'Ongoing Face Recognition Vendor Test Part 2: Identification', National Institute of Standards and Technology. Available online: https://doi.org/10.6028/NIST.IR.8238 (accessed 12 October 2020).

Guattari, Félix. (2000 [1989]), *The Three Ecologies*, trans. Ian Pindar and Paul Sutton, London and New Brunswick, NJ: The Athlone Press.

Hacking, Ian. (1986), 'Making up People', in Thomas C. Heller, Morton Sosna and David E. Wellbery (eds), *Reconstructing Individualism: Autonomy, Individuality, and the Self in Western Thought*, 222–236, Stanford, CA: Stanford University Press.

Hall, Stuart. (2007), 'Encoding, Decoding', in Simon During (ed.), *The Cultural Studies Reader: Third Edition*, 477–87, London: Routledge.

Hansen, Mark. (2003), 'Affect as Medium, or the "Digital-Facial-Image"', *Journal of Visual Culture*, 2 (2): 205–28.

Haraway, Donna. (2016), *Staying with the Trouble: Making Kin in the Chthulucene*, Durham and London: Duke University Press.

Haraway, Donna. (2017), 'Symbiogenesis, Sympoiesis, and Art Science Activisms for Staying with the Trouble', in Anna Tsing, Heather Swanson, Elaine Gan and Nils Bubandt (eds), *Arts of Living on a Damaged Planet: Monsters of the Anthropocene*, M25–M50, Minneapolis: University of Minnesota Press.

Harman, Graham. (2012), *Weird Realism: Lovecraft and Philosophy*, Winchester and Washington: Zero Books.

Hatfield, Charles. (2009), 'An Art of Tensions', in Jeet Heer and Kent Worcester (eds), *A Comics Studies Reader*, 132–48, Jackson: University Press of Mississippi.

Hayles, N. Katherine. (1999), *How We Became Posthuman: Virtual Bodies in Cybernetics, Literature, and Informatics*, Chicago and London: The University of Chicago Press.

Hayles, N. Katherine. (2000), 'Flickering Connectivities in Shelley Jackson's *Patchwork Girl*: The Importance of Media-Specific Analysis', *Postmodern Culture*, 10 (2). Available online: https://muse.jhu.edu/article/27720 (accessed 6 February 2019).

Heinlein, Robert A. (2017 [1956]), *Time for the Stars*, Rockville, MD: ARC Manor.

Hess, Aaron. (2015), 'The Selfie Assemblage', *International Journal of Communication*, 9: 1629–46.

Hirsch, Marianne. (1997), *Family Frames: Photography, Narrative and Postmemory*, Cambridge, MA and London: Harvard University press.

Hofstadter, Douglas. (2007), *I am a Strange Loop*, New York: BasicBooks.

Holl, Ute. 'Trance Techniques, Cinema and Cybernetics', in Heiki Behrend, Anja Dreschke and Martin Zillinger (eds), *Trance Mediums and New Media: Spirit Possession in the Age of Technical Reproduction*, 264–82, New York: Fordham Scholarship Online.

Hörl, Erich. (2017), 'Introduction to General Ecology: The Ecologization of Thinking', trans. Nils F. Schott, in Erich Hörl with James Burton (eds), *General Ecology: The New Ecological Paradigm*, 1–74, London: Bloomsbury.

Hui, Yuk. (2019), *Recursivity and Contingency*, London and New York: Rowman & Littlefield.

Hui, Yuk. (2020), 'Machine and Ecology', *Angelaki*, 25 (4): 54–66.

Huxley, Aldous. (2008 [1932]), *Brave New World*, London: Random House.

Jackson, Shelley. (1995), *Patchwork Girl*, Watertown, MA: Eastgate Systems.

Jackson, Shelley. (1997), 'Stitch Bitch: The Patchwork Girl'. Available online: http://web.mit.edu/m-i-t/articles/jackson.html (accessed 6 February 2019).

Jackson, Shelley. (2016), *Half Life*, London: HarperPerennial.

Jackson, Shelley. (2018), *Riddance: Or: The Sybil Joines Vocational School for Ghost Speakers and Hearing-Mouth Children*, New York: Black Balloon Publishing.

Jackson, Zakiyyah Iman. (2020), *Becoming Human: Matter and Meaning in an Antiblack World*, New York: NYU Press.

Jameson, Fredric. (2007), 'The Space of Science Fiction: Narrative in van Vogt', in Fredric Jameson, (ed.), *Archaeologies of the Future: The Desire Called Utopia and Other Science Fictions*, 314–27, London: Verso.

Jameson, Fredric. (2007), 'After Armageddon: Character Systems in *Dr Bloodmoney*', in Fredric Jameson, (ed.), *Archaeologies of the Future: The Desire Called Utopia and Other Science Fictions*, 349–62, London: Verso.

Jameson, Fredric. (2007), 'History and Salvation in Philip K. Dick', in Fredric Jameson, (ed.), *Archaeologies of the Future: The Desire Called Utopia and Other Science Fictions*, 363–82, London: Verso.

Johnson, Christopher. (1999), 'Ambient Technologies, Uncanny Signs', *Oxford Literary Review*, 21: 117–34.

Kawabata, Yasunari. (2006 [1962]), *The Old Capital*, trans. J. Martin Holman, Berkeley: Counterpoint.

Kawabata, Yasunari. (2019 [1968]), '"Japan, the Beautiful and Myself": Nobel Lecture', in *Dandelions*, trans. Michael Emmerich, 119–32, London: Penguin.

Kay, Lily E. (1997), 'Cybernetics, Information, Life: The Emergence of Scriptural Representations of Heredity', *Configurations*, 5 (1): 23–91.

Khamis, Susie, Lawrence Ang and Raymond Welling. (2017), 'Self-Branding, "Micro-Celebrity" and the Rise of Social Media Influencers', *Celebrity Studies*, 8 (2): 191–208.

Keller, Evelyn Fox. (2015), 'The Postgenomic Genome', in Sarah Richardson and Hallam Stevens (eds), *Postgenomics: Perspectives on Biology after the Genome*, 9–31, Durham, NC: Duke University Press.

Kember, Sarah. (2014), 'Face Recognition and the Emergence of Smart Photography', *Journal of Visual Culture*, 13 (2): 182–99.

King, Stephen. (1977), *The Shining*, London: New English Library.

Kristóf, Ágota. (2014 [1986]), *The Notebook*, trans. Alan Sheridan, London: CB editions.

La Jetée (1962), [Film] Dir. Chris Marker, France: Argos Films.

Lappé, Martine and Hannah Landecker. (2016), 'How the Genome Got a Life Span', *New Genetics and Society*, 34 (2): 152–76.

Lawal, Babatunde. (2011), 'Sustaining Oneness in Their Twoness: Poetics of Twin Figures (Ère Ìbejì) among the Yoruba', in Philip M. Peek (ed.), *Twins in African and Diaspora Cultures: Double Trouble, Twice Blessed*, 81–98, Bloomington and Indianapolis: Indiana University Press.

Lazzarato, Maurizio. (2014), *Signs and Machines: Capitalism and the Production of Subjectivity*, trans. Joshua David Jordan. Los Angeles: Semiotext(e).

Leaver, Tama, Tim Highfield and Crystal Abidin. (2020), *Instagram: Visual Social Media Cultures*, Cambridge: Polity.

Lee-Morrison, Lila. (2019), *Portraits of Automated Facial Recognition: On Machinic Ways of Seeing the Face*, Bielefeld: Transcript Verlag.

Levinas, Emmanuel. (1999), *Alterity and Transcendence*, trans. Michael B. Smith, New York: Columbia University Press.

Lévi-Strauss, Claude. (1995), *The Story of Lynx*, trans. Catherine Tihanyi, Chicago and London: The University of Chicago Press.

Lewis, Alex and Marcus with Joanna Hodgkin. (2013), *Tell Me Who I Am*, London: Hodder & Stoughton.

Lewontin, Richard, Steven Rose and Leon Kamin. (1985), *Not in Our Genes: Biology, Ideology, and Human Nature*, London: Random House.

Looper (2012), [Film] Dir. Rian Johnson, USA: TriStar Pictures.

Lost Highway (1997), [Film] Dir. David Lynch, USA: Asymmetrical Productions.

Lovecraft, H.P. (2012 [1929]), 'The Dunwich Horror', in Leslie S. Klinger (ed.), *The New Annotated H.P. Lovecraft*, 343–87, New York and London: Liveright Publishing Corporation.

Luckhurst, Roger. (2007), *The Invention of Telepathy 1870–1901*, Oxford: Oxford University Press.

Luckhurst, Roger. (2015), 'Diagnosing Dick', in Alexander Dunst and Stefan Schlensag (eds), *The World According to Philip K. Dick*, 13–29, New York: Palgrave Macmillan.

Luckhurst, Roger. (2017), 'The Weird: A Dis/Orientation', *Textual Practice*, 31 (6): 1041–61.

Lury, Celia. (2009), 'Brand as Assemblage', *Journal of Cultural Economy*, 2 (1–2): 67–82.

Manning, Erin. (2013), *Always More Than One: Individuation's Dance*, Durham, NC: Duke University Press.

Madrigal, Alexis C. (2018), 'No, You Don't Really Look Like that', *The Atlantic*, 16 December. Available online: https://www.theatlantic.com/technology/archive/2018/12/your-iphone-selfies-dont-look-like-your-face/578353/?utm_source=facebook&utm_campaign=the-atlantic&utm_content=edit-promo&utm_term=2018-12-18T20%3A54%3A27&utm_medium=social (accessed 5 July 2021).

Mafe, Diana Adesola. (2012), 'Ghostly Girls in the "Eerie Bush": Helen Oyeyemi's *The Icarus Girl* as Postcolonial Female Gothic Fiction', *Research in African Literatures*, 43 (3): 21–35.

Marche, Stephen. (2018), 'Modern Family', *Times Literary Supplement*, 22 August.

Martell, Joanne. (2000), *Millie-Christine: Fearfully and Wonderfully Made*, Winston-Salem, NC: John F. Blair Publishing.

Mashable. (2017), 'Is the iPhone X's Facial Recognition Twin Compatible?', *YouTube*, 31 October. Available online: https://www.youtube.com/watch?v=e8-yupM-6Oc (accessed 7 July 2021).

Matrix Reloaded (2003), [Film] Dir. Lan and Lilly Wachowski, USA: Warner Bros.

McCormack, Tom. (2011), 'Code Eroded: At GLI.TC/H 2010, RHIZOME', in Nick Briz, Evan Meaney, Rosa Menkman, William Robertson, Jon Satrom and Jessica Westbrook (eds), *GLI.TC/H 20111: READER[ROR]*, 15–19, Chicago: Unsorted Books.

McHale, Brian. (1987), *Postmodernist Fiction*, London: Routledge.

Miéville, China. (2012), *Embassytown*, London: Pan.

Minevich, Mark. (2021), 'Digital Twin Technology May Save Lives and Solve the Biggest Challenges Facing Humanity', *The Hill*, May 22. Available online: https://thehill.com/opinion/technology/554881-digital-twin-technology-may-save-lives-and-solve-the-biggest-challenges (accessed 29 June 2021).

Minority Report (2002), [Film] Dir. Steven Spielberg, USA: Twentieth Century.

Mitchell, W. J. T. (2008), 'Addressing Media', *MediaTropes eJournal*, 1: 1–18.

Mitchell, W. J. T. (2012), *Seeing Through Race*, Cambridge, MA: Harvard University Press.

Moench, Doug and Alex Niño. (1978), *Heavy Metal Presents Theodore Sturgeon's More Than Human: The Graphic Story Version*, New York: Byron Press Visual Publications.

Morton, Timothy. (2016), *Dark Ecology: For a Logic of Future Coexistence*, New York: Columbia University Press.

Moten, Fred. (2003), *In the Break: The Aesthetics of the Black Radical Tradition*, Minneapolis and London: University of Minnesota Press.

Myers, Frederic W.H. (1892), 'The Subliminal Consciousness. Chapter 1: General Characteristic and Subliminal Messages', *Proceedings of the Society for Psychical Research*, 7: 298–327.

Myers, Frederic W.H. (1903), *Human Personality and Its Survival of Bodily Death, Volume I*, London: Longmans, Green, and Co.

Nabokov, Vladimir. (1971 [1958]), 'Scenes from the Life of a Double Monster', in *Nabokov's Dozen: Thirteen Stories*, London: Penguin Books.

Neyrat, Frédéric, (2018), *The Unconstructable Earth: An Ecology of Separation*, New York: Fordham University Press.

Ngai, Sianne. (2005), *Ugly Feelings*, Cambridge, MA and London: Harvard University Press.

Niffenegger, Audrey. (2010), *Her Fearful Symmetry*, London: Vintage.

A Nightmare on Elm Street (1984), [Film] Dir Wes Craven, USA: New Line Cinema.

de Nooy, Juliana and Bronwyn Statham, (1998), 'Telling the Good from the Bad in Twin Films', *Journal of Media & Cultural Studies*, 12 (3): 279–93.

Orphan Black (2013–17), [TV programme] Canada: Temple Street Productions.

Owens, Jay. (2018), 'The Age of Post-Authenticity and the Ironic Truths of Meme Culture,' *Medium*, April 11. Available online: https://medium.com/s/story/post-0.authenticity-and-the-real-truths-of-meme-culture-f98b24d645a0 (accessed 15 November 2018).

Oyeyemi, Helen. (2005), *The Icarus Girl*, London: Bloomsbury.

The Parent Trap (1961), [Film] Dir. David Swift, USA: Walt Disney Productions.
The Parent Trap (1998), [Film] Dir. Nancy Meyers, USA: Walt Disney Pictures.
Parikka, Jussi. (2016), *Digital Contagions: A Media Archaeology of Computer Viruses: Second Edition*, New York: Peter Lang.
Pausanias. (1935), *Description of Greece, Volume IV: Books 8.22-10* (Arcadia, Boetia, Phocis and Ozolian Locri), trans. W. H. S. Jones. Loeb Classical Library 297, Cambridge, MA: Harvard University Press.
Peek, Philip M. (2011), 'Introduction: Beginning to Rethink Twins', in Philip M. Peek (ed.), *Twins in African and Diaspora Cultures: Double Trouble, Twice Blessed*, 1–36, Bloomington and Indianapolis: Indiana University Press.
Personal Shopper (2016), [Film] Dir. Olivier Assayas, France: CG Cinéma.
Pickering, Andrew. (2010), *The Cybernetic Brain: Sketches of Another Future*, Chicago and London: University of Chicago Press.
Pifer, Ellen. (1981), 'Locating the Monster in Nabokov's "Scenes from the Life of a Double Monster"', *Studies in American Fiction*, 9 (1): 97–101.
Piontelli, Alessandra. (2008), *Twins in the World: The Legends They Inspire and the Lives They Lead*, New York: Palgrave Macmillan.
Playfair, Guy Lyon. (2012), *Twin Telepathy*, Guildford: White Crow Books.
Plomin, Robert. (2018), *Blueprint: How DNA Makes Us Who We Are*, London: Allen Lane.
Poe, Edgar Allan. (2005 [1839]), 'The Fall of the House of Usher', in Graham Clarke (ed.), *Tales*, 137–55, Stroud: Nonsuch.
Poe, Edgar Allan. (2005 [1839]), 'William Wilson', in Graham Clarke (ed.), *Tales*, 1–20, Stroud: Nonsuch.
Posadas, Baryon Tensor. (2009), 'Rampo's Repetitions: The Doppelganger in Edogawa Rampo and Tsukamoto Shin-ya', *Japan Forum*, 21 (2): 161–82.
du Preez, Derek. (2021), 'How Rolls-Royce Is Improving Engine Sustainability with Real-Time Data and Digital Twins', *diginomica*, April 5. Available online: https://diginomica.com/how-rolls-royce-improving-engine-sustainability-real-time-data-and-digital-twins (accessed 29 June 2021).
Punday, David. (2004), 'Involvement, Interruption, and Inevitability: Melancholy as an Aesthetic Principle in Game Narratives', *SubStance*, 33 (3): 80–107.
Rampo, Edogawa. (2012), *Japanese Tales of Mystery and Imagination*, trans. James B. Harris, Tokyo: Tuttle Publishing.
Rank, Otto. (1971), *The Double: A Psychoanalytic Study*, trans. Harry Tucker, Jr., Chapel Hill: University of North Carolina Press.
Renne, Elisha P. and Misty L. Bastian, (2001), 'Reviewing Twinship in Africa', *Ethnology*, 40 (1): 1–11.
Ring (1998), [Film] Dir. Hideo Nakata, Japan: Basara Pictures.
Rollefson, J. Griffith. (2008), 'The "Robot Voodoo Power" Thesis: Afrofuturism and Anti-Anti-Essentialism from Sun Rato Kool Keith', *Black Music Research Journal*, 28 (1): 83–109.

Rose, Nikolas. (2007), *The Politics of Life Itself: Biomedicine, Power, and Subjectivity in the Twenty-First Century*, Princeton and Oxford: Princeton University Press.

Royle, Nicholas. (2003), *The Uncanny*, Manchester: Manchester University Press.

Rubinstein, Daniel and Katrina Sluis. (2013), 'The Digital Image in Photographic Culture: Algorithmic Photography and the Crisis of Representation', in Martin Lister (ed.), *The Photographic Image in Digital Culture: Second Edition*, 22–40, London and New York: Routledge.

Savat, David. (2013), *Uncoding the Digital: Technology, Subjectivity and Action in the Control Society*, New York: Palgrave MacMillan.

Schwartz, Hillel. (1996), *The Culture of the Copy: Striking Likenesses, Unseasonable Facsimiles*, New York: Zone Books.

Sconce, Jeffrey. (2000), *Haunted Media: Electronic Presence from Telegraphy to Television*. Durham, NC: Duke University Press.

Seconds Apart (2011), [Film] Dir. Antonio Negret, USA: After Dark Films.

Segal, Nancy. (2012), *Born Together – Reared Apart: The Landmark Minnesota Twin Study*. Cambridge, MA and London: Harvard University Press.

Sekula, Allan. (1986), 'The Body and the Archive', *October* 39: 3–64.

Seymour, Nicole. (2017), *Strange Natures: Futurity, Empathy, and the Queer Ecological Imagination*, Champaign: Illinois Scholarship Online.

Shaviro, Steven. (2017), 'Review of *Rosewater* (2016) by Tade Thompson', *The Pinocchio Theory*, January 16. Available online: http://www.shaviro.com/Blog/?p=1422 (accessed 29 June 2021).

The Shining (1980), [Film] Dir. Stanley Kubrick, USA: Warner Bros.

Simondon, Gilbert. (1989), *L'Individuation Psychique et Collective*, Paris: Aubier.

Sisters (1972), [Film] Dir. Brian de Palma, USA: Pressman-Williams Enterprises.

Sloterdijk, Peter. (2011), *Spheres Volume I: Bubbles: Microspherology*, trans. Wieland Hoba, Los Angeles, CA: Semiotext(e).

Stam, Robert. (1997), 'From Hybridity to the Aesthetics of Garbage', *Social Identities*, 3 (2): 275–90.

Stoker, Bram. (2016 [1886]), *The Dualitists; or the Death Doom of the Double Born*, Auckland: The Floating Press.

Sturgeon, Theodore. (2000 [1953]), *More than Human*, London: Orion

Sutin, Lawrence. (2005), *Divine Invasions: A Life of Philip K. Dick*, New York: Carroll & Graf Publishers.

Tech Insider. (2017), 'We Put the iPhone X's Face ID to the Test with Twins', *YouTube*, 31 October. Available online https://www.youtube.com/watch?v=rFoSPZBHsmE (accessed 7 July 2021).

Tell Me Who I Am (2019), [Film] Dir. Ed Perkins, UK: Lightbox.

Terranova, Tiziana. (2004), *Network Culture: Politics for the Information Age*, London: Pluto Press.

Thacker, Eugene. (2005), *The Global Genome: Biotechnology, Politics, and Culture*, Cambridge, MA and London: The MIT Press.

Thacker, Eugene. (2014), 'Dark Media', in Alexander R. Gallway, Eugene Thacker and McKenzie Wark (eds), *Excommunication: Three Inquiries in Media and Mediation*, 77–149, Chicago and London: The University of Chicago Press.

Than, Ker. (2016), 'A Brief History of Twin Studies', *Smithsonian Magazine*, March 4. Available online: https://www.smithsonianmag.com/science-nature/brief-history-twin-studies-180958281/ (accessed 11 May 2021).

Thompson, Tade. (2019), *The Rosewater Insurrection*, London: Orbit.

Tsing, Anna Lowenhaupt. (2015), *The Mushroom at the End of the World: On the Possibility of Life in Capitalist Ruins*, Princeton and Oxford: Princeton University Press.

Turner, Victor. (1969), *The Ritual Process: Structure and Anti-Structure*, New York: Aldine de Gruyter.

Twinsters (2015), [Film] Dir. Samantha Futerman and Ryan Miyamoto, USA: Small Package Films.

The Unborn (2009), [Film] Dir. David S. Goyer, USA: Rogue Pictures.

van Veen, tobias c. (2013), 'Vessels of Transfer: Allegories of Afrofuturism in Jeff Mills and Janelle Monáe', *Dancecult: Journal of Electronic Dance Music Culture*, 5 (2): 7–41.

van Veen, tobias c. and Reynaldo Anderson. (2018), 'Future Movements: Black Lives, Black Politics, Black Futures: An Introduction', *TOPIA: Canadian Journal of Cultural Studies*, 39: 5–21.

van Vogt, A.E. (1960 [1940]), *Slan*, London: Panther Books.

van Vogt, A. E. (2013 [1943]), *The Weapon Makers*, Los Angeles: Agency Editions.

Vanderhaeghe, Stéphane. (2010), 'How to Unread Shelley Jackson', *Transatlantica*, 2: 1–10.

Vandermeer, Jeff. (2014), *Acceptance*. London: Fourth Estate.

Viney, William. (2013), 'The Significance of Twins in the Middle Ages.' Available online: https://thewonderoftwins.wordpress.com/2013/07/23/the-significance-of-twins-in-medieval-and-early-modern-europe/ (accessed 29 June 2021).

Viney William. (2015), 'Anthropology's Twins.' Available online: https://thewonderoftwins.wordpress.com/tag/levi-strauss/ (accessed 29 June 2021).

Viney, William. (2016), 'Getting the Measure of Twins,' in Anne Whitehead, Angela Woods, Sarah Atkinson, Jane Macnaughton and Jennifer Richards (eds), *The Edinburgh Companion to the Critical Medical Humanities*, 104–19, Edinburgh: Edinburgh University Press.

Viney, William. (2018), 'Experimenting in the Biosocial: The Strange Case of Twin Research', in Maurizio Meloni, John Cromby, Des Fitzgerald and Stephanie Lloyd (eds), *The Palgrave Handbook of Biology and Society*, 143–66, New York: Palgrave.

Viney, William. (2021), *Twins: Superstitions and Marvels, Fantasies and Experiments*, London: Reaktion Books.

Viveiros de Castro, Eduardo. (2014), *Cannibal Metaphysics: For a Post-Structural Anthropology*, trans. and ed. Peter Skafish, Minneapolis, MN: Univocal.

Wahrman, Dror. (2004), *The Making of the Modern Self: Identity and Culture in Eighteenth-Century England*, New Haven and London: Yale University Press.

Wallace, Marjorie. (1996), *The Silent Twins*, London: Vintage.

Walter, W. Grey. (1961 [1953]), *The Living Brain*, London: Penguin Books.

Wark, McKenzie. (2017), 'Eduardo Viveiros de Castro: In and Against the Human', *Verso blog*, June 12. Available online: https://www.versobooks.com/blogs/3265-eduardo-viveiros-de-castro-in-and-against-the-human (accessed 29 June 2021).

Wasson, Sara. (2004), 'Love in the Time of Cloning: Science Fictions of Transgressive Kinship', *Extrapolation*, 45 (2): 130–44.

Welch, Patricia. (2012), 'Excess, Alienation and Ambivalence: Edogawa Rampo's Tales of Mystery and Imagination', in *Japanese Tales of Mystery and Imagination*, trans. James B. Harris, 13–25, Tokyo: Tuttle Publishing, 2012.

Wiener, Norbert. (1989 [1950]), *The Human Use of Human Beings: Cybernetics and Society*, London: Free Association Books.

Wilkinson, Alissa. (2014), 'What's with All the Movies about Doppelgängers?' *The Atlantic*, March 14. Available online: https://www.theatlantic.com/entertainment/archive/2014/03/whats-with-all-the-movies-about-doppelg-ngers/284413/ (accessed 10 May 2021).

Winnicott, D.W. (1991 [1971]), *Playing and Reality*, London: Routledge.

Wright, Lawrence. (1997), *Twins: Genes, Environment and the Mystery of Identity*, London: Weidenfeld & Nicolson.

Yusoff, Kathryn. (2018), *A Billion Black Anthropocenes or None*, Minneapolis: The MIT Press.

A Zed & Two Noughts (1985), Dir. Peter Greenaway, London: BFI.

Zunshine, Lisa. (2008), *Strange Concepts and the Stories They Make Possible*, Baltimore: The Johns Hopkins University Press.

Zhang, Sarah. (2018), 'Your DNA Is Not Your Culture', *The Atlantic*, September 25. Available online: https://www.theatlantic.com/science/archive/2018/09/your-dna-is-not-your-culture/571150/ (accessed 11 May 2021).

Index

6LACK 157–8
26a (Evans) 144
abiku 144–5, 149
Alaimo, Staicy 125
allegory 141–2, 151, 157
Amerindian perspectivalism 116–19
anthropology 116–19
Arbus, Diane 73, 125, 188–91

Bakker, Gerbrand 113
Barnum, P.T. 152
Bertillon, Alphonse 170
Big Little Lies 181
Bight of the Twin, The 161–3
biometric systems 170
biopolitics 13–18, 27 n.49, 101–2
Blas, Zach 182
Boklage, Charles 99
Bollmer, Grant 4, 17
Bouchard, Thomas 11, 13
Bourgeois, Larry and Laurant 154–60
Brave New World (Huxley) 6, 40
Brooklyn and Bailey 2–4
Bunker, Chang and Eng 5, 152

Chaudhuri, Uni 112
Cheney-Lippold, John 16–17
Chiew, Florence and Alison Barnwell 22, 133, 200
chimerism 99
Chun, Wendy 20, 175–6
Clark, Timothy 125
Clermont Twins 187
clones 6, 40–1, 85–6
Colwell, C. 16
conjoined twins 5, 26 n.14, 31–5, 100, 127–32, 152–3, 185
control. *See* Deleuze, Gilles
Corsican Brothers, The (Dumas) 31

Danowski, Débora and Eduardo Viveiros de Castro 111–12, 115–16
datafication 173–4
Dead Ringers 97

Deleuze, Gilles
    death drive 104
    Difference and Repetition 104
    disjunctive synthesis 24
    dividuality 16
    'Postscript on the Societies of Control' 16, 175
Deleuze, Gilles and Félix Guattari
    dualities 118–19
    faciality 186
    machinic enslavement 52–3
dem (Kelley) 141
Dick, Philip K.
    Do Androids Dream of Electric Sheep? 65–6
    Dr Bloodmoney 59
    Exegesis of Philip K. Dick, The 61–4
    Flow My Tears, the Policeman Said 58–9
    relationship with sister Jane 55
    Simulacra, The 58
    Three Stigmata of Palmer Eldritch, The 57
    Ubik 57
    Valis 61–4
digital twins 18, 197–9
Dillon, Karen 141
Dolan Twins 2, 169, 171–3
doppelgängers. *See* doubles
double consciousness 156–7
doubles 2, 13, 31
'Dualists, The' (Stoker) 79–80
Dubois, W.E.B. 156–7
Dumas, Alexandre 31
'Dunwich Horror, The' (Lovecraft) 81–5

early modern theatre 3, 6, 88–9
Edgar Allan Poe
    'Fall of the House of Usher, The' 76–9
    'William Wilson' 78
Edugyan, Esi 144, 160–1
Elmer and Elsie 36–8, 40–1, 67 n.30
Emecheta, Buchi 161
Enchi, Fumiko 90–4
Engelstein, Stefani 18–20

ere ibeji 144, 162
eugenics 13, 35, 39, 43, 46, 97, 102, 125, 141
Evans, Diana 144–5, 149–52
Evans-Pritchard, E.E. 117

facial recognition technology 169–76
fetus-in-fetu 59–60, 123
Fisher, Mark 126
von Foerster, Heinz 62–3
Foucault, Michel 16, 173
Frazer, James George 111
Freaks 185
Freud, Sigmund 13
   Beyond the Please Principle 103
   death drive 103–4
   'Ego and the Id, The' 103
   'Mourning and Melancholia' 93, 101
   'Uncanny, The' 73, 80
Frost, Samantha 10, 124

Galton, Francis 6, 8–9, 170
game theory 64
genetics research 6, 8–9, 12, 14, 96, 124, 198
glitch 143, 157–9, 174, 182, 198–200
Gomel, Elana and Stephen Weninger 101–2
gothic fiction 6, 13, 31, 41, 45, 73, 76–81, 84, 90, 94, 101, 128, 130, 144
   See also doubles
Grey Walter, William 36–8, 40–1, 67 n.30
Grosz, Elizabeth 185–6
Guattari, Félix 34, 51, 67 n.21

Hal, Stuart 12
Haraway, Donna 112–13
Harman, Graham 83
Hayles, Katherine 32, 56, 57–9, 131–2
Her Fearful Symmetry (Niffenegger) 90, 179
Hilton, Violet and Daisy 185
Hirsch, Marianne 180–1
homophily 20, 175–6
Hopkinson, Nalo 151
Hörl, Erich 34
Hui, Yuk 7, 21, 112, 122, 132–3, 198
Huxley, Aldous 6, 40

Icarus Girl, The (Oyeyemi) 143–7
'Identical Twins, Roselle, New Jersey'. *See* Arbus, Diane
in vitro fertilization (IVF) 9, 19–20

influencers. *See* social media stardom
IVF. *See* in vitro fertilization (IVF)

Jackson, Shelley
   Half Life 127–31
   Patchwork Girl 131–2
Jackson, Zakiyyah Iman 150–1, 153

Kawabata, Yasunari 119–23
Kehinde (Emecheta) 161
Keller, Evelyn Fox 11
Kelley, William Melvin 141
kinship 18–20, 117, 141–2, 158
Kristóf, Ágota 113

Lazzarato, Mauricio 52
Lévi-Strauss, Claude
   myth 57
   Story of Lynx, The 22–3, 116–18
   View from Afar, The 114
Lost Highway 81, 127
Lovecraft, H.P. 81–5

masks 179–2
Masks (Enchi) 90–4
Matrix Reloaded, The 156
Maxwell's Demon 62–3
McCoy, Millie and Christine 152–4
Meatyard, Ralph Eugene 179–80
Men in Black: International 156
Minnesota Study of Twins Reared Apart, The 11, 13–15, 22, 100, 127
More Than Human (Moench and Niño) 54
More Than Human (Sturgeon) 49–54
Morrison, Toni 141, 146
Myers, Frederic 32, 45

Nabokov, Vladimir 31–5
Narcissus 37, 41
Neyrat, Frédéric 123
Notebook, The (Kristóf) 113

Old Capital, The (Kawabata) 119–23
Oyeyemi, Helen 143–7

Paradise (Morrison) 141, 146
Parent Trap 1
Parikka, Jussi 171
performance 152–60
Personal Shopper 75–6

Index

Peskine, Alexis 139–40
photography
    biometric photography 88, 169–76
    digital images 174–6
    family photography 179–81
    memory and photography 95
    portraiture 124, 179
    selfies 183–8
    spiritualist photography 90
Pickering, Andrew 38
Piontelli, Alessandra 135 n.33
P-Orridge, Breyer 161–3
posthumanism 4, 33, 172
psychotherapy 50–1
    See also Freud, Sigmund
Pudd'nhead Willson (Twain) 141

Rampo, Edogawa 87–9
Rank, Otto 84
recursion 2, 7, 20, 21, 62, 112, 122, 159
reuniting twins 1–2
Romulus and Remus 148
Rose, Nikolas 8
Rosewater Insurrection, The (Thompson) 147–52
Royle, Nicholas 160

'Scenes from the Life of a Double Monster' (Nabokov) 31–5
Schwartz, Hillel 5–6, 21–2, 193 n.36, 198
Second Life of Samuel Tyne, The (Edugyan) 144, 160–1
Seconds Apart 105–6 n.17
Segal, Nancy 11, 127
Shannon, Claude 75
Shining, The 73, 126–7, 188
Siamese twins. *See* conjoined twins
siblings 18–20
Siemens, Hermann 8
Silent Twins, The (Wallace) 147, 160–1
Sisters 108–9 n.90
social media
    celebrity 2–5, 17, 171–4, 178, 183, 187
    connectivity 1–5
Society for Psychical Research (SPR) 31–2, 45
Southern Reach Trilogy, The (Vandermeer) 85–6
Stiegler, Bernard 59–60
Stoker, Bram 79–80

Sturgeon, Theodore 49–54
superfecundation 141

Tell Me Who I Am 94–6
temporality. *See* time
Terranova, Tiziana 17
Thacker, Eugene 14, 75, 89–90, 131
Third Twin, The (Follet) 15–16
Thompson, Tade 147–52
Three Identical Strangers 39
time 123–32
Time for the Stars (Heinlein) 43–5
time travel 136 n.74
town twinning 22–3
transitional objects 59–60
Tsing, Anna 6, 125–6
Turner, Victor 5, 117
Twain, Mark 141
Twelfth Night (Shakespeare) 3
Twin, The (Bakker) 113
twin method. *See* genetics research
twin paradox (Paul Langevin) 43
twin studies. *See* genetics research
Twins, Les. *See* Bourgeois, Larry and Laurant
'Twins, The' (Rampo) 87–9
Twinsters 1–2
twinstrangers.net 177–8

Unborn, The 97–8
uncanny 13, 36–7, 41, 45, 73, 76–81, 84–5, 87–9, 144, 160–1

van Vogt, A.E.
    Slan 42
    Weapon Makers, The 45–9
Vandermeer, Jeff 85–6
vanishing twins 6
Viney, William 5, 9, 19–20, 100, 117, 170
Viveiros de Castro, Eduardo 21, 116–19

Wallace, Marjorie 147, 160–1
Wiener, Norbert 33, 38–40, 50, 64
Winnicott, Donald 59–60
Wright, Lawrence 11, 13, 15, 123
Wyndham, John 49

Zed & Two Noughts, A 101–4
Zelewski, Peter 179
Zunshine, Lisa 3–4, 18, 88–9, 169, 185

Printed in the USA
CPSIA information can be obtained
at www.ICGtesting.com
LVHW011627120724
785347LV00001B/70